Use R!

Advisors:
Robert Gentleman · Kurt Hornik · Giovanni Parmigiani

For other titles published in this series, go to
http://www.springer.com/series/6991

Christian P. Robert · George Casella

Introducing Monte Carlo Methods with R

 Springer

Christian P. Robert
Université Paris Dauphine
UMR CNRS 7534
CEREMADE
place du Maréchal de Lattre
de Tassigny
75775 Paris cedex 16
France
xian@ceremade.dauphine.fr

George Casella
Department of Statistics
University of Florida
103 Griffin-Floyd Hall
Gainesville FL 32611-8545
USA
casella@stat.ufl.edu

Series Editors
Robert Gentleman
Department of Bioinformatics
  and Computational Biology
Genentech South San Francisco
CA 94080
USA

Kurt Hornik
Department of Statistik and Mathematik
Wirtshchaftsuniversität Wien Augasse 2-6
A-1090 Wien
Austria

Giovanni Parmigiani
Department of Biostatistics
  and Computational Biology
Dana-Farber Cancer Institute
44 Binney Street
Boston, MA 02115
USA

ISBN 978-1-4419-1575-7   e-ISBN 978-1-4419-1576-4
DOI 10.1007/978-1-4419-1576-4
Springer New York Dordrecht Heidelberg London

Library of Congress Control Number: 2009941076

Springer is part of Springer Science+Business Media (www.springer.com)

To our parents, who taught us much in many ways.

"What if you haven't the data?"
"Then we shall proceed directly to the brandy and cigars."

Lyndsay Faye
The Case of Colonel Warbuton's Madness

Preface

"After that, it was down to attitude."

Ian Rankin
Black & Blue

The purpose of this book is to provide a self-contained entry into Monte Carlo computational techniques. First and foremost, it must not be confused with a programming addendum to our earlier book *Monte Carlo Statistical Methods* whose second edition came out in 2004. The current book has a different purpose, namely to make a general audience familiar with the programming aspects of Monte Carlo methodology through practical implementation. Not only have we introduced R at the core of this book, but the emphasis and contents have changed drastically from *Monte Carlo Statistical Methods*, even though the overall vision remains the same. Theoretical foundations are intentionally avoided in the current book.

Indeed, the emphasis on *practice* is a major feature of this book in that its primary audience consists of graduate students in statistics, biostatistics, engineering, etc., who need to learn how to use simulation methods as a tool to analyze their experiments and/or datasets. The book should appeal to scientists in all fields, given the versatility of these Monte Carlo tools. It can also be used for a more classical statistics audience when aimed at teaching a quick entry into modern computational methods based on R, at the end of an undergraduate program for example, even though this may prove challenging for some students.

The choice of the programming language R, as opposed to faster alternatives such as Matlab or C and more structured constructs such as BUGS, is due to its pedagogical simplicity and its versatility. Readers can easily conduct experiments in their own programming language by translating the examples provided in this book. (Obviously, the book can also supplement other textbooks on Bayesian modeling at the graduate level, including our books *Bayesian Choice* (Robert, 2001) and *Monte Carlo Statistical Methods* (Robert and Casella, 2004).) This book can also be viewed as a companion to, rather than a competitor of, Jim Albert's Use R! book *Bayesian Computation with R* (Albert, 2009). Indeed, taken as a pair, these two books can provide a fairly thorough introduction to Monte Carlo methods and Bayesian modeling.

We stress that, at a production level (that is, when using advanced Monte Carlo techniques or analyzing large datasets), R cannot be recommended as the default language, but the expertise gained from this book should make the switch to another language seamless.

Contrary to usual practice, many exercises are interspersed within the chapters rather than postponed until the end of each chapter. There are two reasons for this stylistic choice. First, the results or developments contained in those exercises are often relevant for upcoming points in the chapter. Second, they signal to the student (or to any reader) that some pondering over the previous pages may be useful before moving to the following topic and so may act as self-checking gateways. Additional exercises are found at the end of each chapter, with abridged solutions of the odd-numbered exercises provided on our Webpages as well as Springer's.

Thanks

We are immensely grateful to colleagues and friends for their help with this book, in particular to the following people: Ed George for his comments on the general purpose of the book and some exercises in particular; Jim Hobert and Fernando Quintana for helpful discussions on the Monte Carlo EM; Alessandra Iacobucci for signaling in due time a fatal blunder; Jean-Michel Marin for letting us recycle the first chapter of *Bayesian Core* (Marin and Robert, 2007) into our introduction to R and for numerous pieces of R advice, as well as pedagogical suggestions; Antonietta Mira for pointing out mistakes during a session of an MCMC conference in Warwick; François Perron for inviting CPR to Montréal and thus providing him with a time window to complete Chapter 8 (only shortened by an ice-climbing afternoon in Québéc!), and also François Perron and Clémentine Trimont for testing the whole book from the perspectives of a professor and a student, respectively; Martyn Plummer for answering queries about coda; Jeff Rosenthal for very helpful exchanges on amcmc; Dimitris Rizopoulos for providing Exercise 7.21; and Phil Spector from Berkeley for making available his detailed and helpful notes and slides on R and C, now partly published as Spector (2009). Comments from both reviewers were especially useful in finalizing the book. We are also grateful to John Kimmel of Springer for his advice and efficiency, as well as for creating the visionary Use R! series and supporting the development of the R language that way. From a distance, we also wish to thank Professors Gentleman and Ihaka for creating the R language and for doing it in open-source, as well as the entire R community for contributing endlessly to its development.

Sceaux and Gainesville   **Christian P. Robert and George Casella**
October 18, 2009

Contents

Preface vii

List of Figures xiii

List of Examples xvii

1   Basic R Programming 1
    1.1 Introduction 2
    1.2 Getting started... 3
    1.3 R objects 5
        1.3.1 The vector class 6
        1.3.2 The matrix, array, and factor classes 9
        1.3.3 The list and data.frame classes 12
    1.4 Probability distributions in R 14
    1.5 Basic and not-so-basic statistics 14
    1.6 Graphical facilities..................................... 26
    1.7 Writing new R functions 31
    1.8 Input and output in R................................... 35
    1.9 Administration of R objects 36
    1.10 The mcsm package....................................... 36
    1.11 Additional exercises.................................... 37

2   Random Variable Generation 41
    2.1 Introduction 42
        2.1.1 Uniform simulation 42
        2.1.2 The inverse transform............................. 44
    2.2 General transformation methods.......................... 46
        2.2.1 A normal generator............................... 47
        2.2.2 Discrete distributions 48
        2.2.3 Mixture representations 50
    2.3 Accept–reject methods 51

      2.4   Additional exercises....................................... 57

3   Monte Carlo Integration.................................. 61
      3.1   Introduction 62
      3.2   Classical Monte Carlo integration....................... 65
      3.3   Importance sampling..................................... 69
            3.3.1   An arbitrary change of reference measure............. 69
            3.3.2   Sampling importance resampling 75
            3.3.3   Selection of the importance function 78
      3.4   Additional exercises.................................... 86

4   Controlling and Accelerating Convergence 89
      4.1   Introduction 90
      4.2   Monitoring variation 91
      4.3   Asymptotic variance of importance sampling estimators 92
      4.4   Effective sample size and perplexity....................... 98
      4.5   Simultaneous monitoring 100
      4.6   Rao–Blackwellization and deconditioning 107
      4.7   Acceleration methods 111
            4.7.1   Correlated simulations 111
            4.7.2   Antithetic variables 113
            4.7.3   Control variates 116
      4.8   Additional exercises....................................122

5   Monte Carlo Optimization 125
      5.1   Introduction 126
      5.2   Numerical optimization methods 127
      5.3   Stochastic search 130
            5.3.1   A basic solution 130
            5.3.2   Stochastic gradient methods 136
            5.3.3   Simulated annealing 140
      5.4   Stochastic approximation 146
            5.4.1   Optimizing Monte Carlo approximations 146
            5.4.2   Missing-data models and demarginalization 150
            5.4.3   The EM algorithm 152
            5.4.4   Monte Carlo EM 157
      5.5   Additional exercises....................................163

6   Metropolis–Hastings Algorithms 167
      6.1   Introduction 168
      6.2   A peek at Markov chain theory...........................168
      6.3   Basic Metropolis–Hastings algorithms 170
            6.3.1   A generic Markov chain Monte Carlo algorithm 171
            6.3.2   The independent Metropolis–Hastings algorithm 175
      6.4   A selection of candidates 182

	6.4.1	Random walks 182
	6.4.2	Alternative candidates 185
6.5	Acceptance rates 192	
6.6	Additional exercises.................................... 195	

7 Gibbs Samplers... 199
7.1 Introduction 200
7.2 The two-stage Gibbs sampler 200
7.3 The multistage Gibbs sampler.......................... 206
7.4 Missing data and latent variables 209
7.5 Hierarchical structures 221
7.6 Other considerations 224
    7.6.1 Reparameterization................................. 224
    7.6.2 Rao–Blackwellization 227
    7.6.3 Metropolis within Gibbs and hybrid strategies 230
    7.6.4 Improper priors 232
7.7 Additional exercises.................................... 234

8 Monitoring and Adaptation for MCMC Algorithms........ 237
8.1 Introduction 238
8.2 Monitoring what and why 238
    8.2.1 Convergence to the stationary distribution 238
    8.2.2 Convergence of averages 240
    8.2.3 Approximating iid sampling 240
    8.2.4 The coda package 241
8.3 Monitoring convergence to stationarity 242
    8.3.1 Graphical diagnoses 242
    8.3.2 Nonparametric tests of stationarity 243
    8.3.3 Spectral analysis 247
8.4 Monitoring convergence of averages 250
    8.4.1 Graphical diagnoses 250
    8.4.2 Within and between variances 253
    8.4.3 Effective sample size 255
    8.4.4 Fixed-width batch means........................... 257
8.5 Adaptive MCMC 258
    8.5.1 Cautions about adaptation 258
    8.5.2 The amcmc package 264
8.6 Additional exercises.................................... 267

References.. 269

Index of R Terms 275

Index of Subjects 279

List of Figures

1.1 Illustrations of the processing of vectors in R. 8
1.2 Illustrations of the processing of matrices in R. 11
1.3 Illustrations of the factor class. 12
1.4 Chosen features of the list class. 13
1.5 Definition of a data frame. 14
1.6 Loess and natural splines 19
1.7 Autocorrelation and partial autocorrelation plots 23
1.8 Simple illustration of bootstrap 24
1.9 Bootstrap linear regression 26
1.10 Spline approximation of monthly deaths 29
1.11 Cumsum illustration for an AR(1) 30
1.12 Range of Brownian motions with confidence band 32
1.13 Some artificial loops in R. 34

2.1 Representation of a uniform random sample 43
2.2 Representations of an exponential random sample 45
2.3 Representation of a binomial random sample 51
2.4 Generation of beta random variables 54

3.1 Evaluation of integrate 63
3.2 Comparison of integrate and area 64
3.3 One-dimensional Monte Carlo integration 67
3.4 Importance sampling approximation of a normal tail 72
3.5 Representation of the posterior $\pi(\alpha, \beta|x)$ 74
3.6 Analysis of a sample from $\pi(\alpha, \beta|x)$ 78
3.7 Infinite variance importance sampler 80
3.8 Convergence of two estimators of the integral (3.9) 84
3.9 Posterior of the regression parameter (β_1, β_2) 86

4.1 Confidence bands for a simple example 92
4.2 Range and confidence for the Cauchy-normal problem (1) 97

4.3   Range and confidence for the Cauchy-Normal problem (2) 98
4.4   ESS and perplexity for the Cauchy-Normal problem 101
4.5   ESS and perplexity for the Cauchy-Normal problem 104
4.6   Brownian confidence band for the Cauchy-Normal problem 106
4.7   Convergence of estimators of $\mathbb{E}\{\exp(-X^2)\}$ 109
4.8   Approximate risks of truncated James–Stein estimators 114
4.9   Impact of dyadic average 116
4.10 Impact of control variates 119
4.11 Impact of control variates in logistic regression 121

5.1   Sequences of MLEs 128
5.2   Newton–Raphson sequences for a mixture likelihood 129
5.3   Simple Monte Carlo maximization 132
5.4   Two Cauchy likelihood maximizations 133
5.5   A Cauchy likelihood approximation 134
5.6   Representation of the function of Example 5.6 136
5.7   Stochastic gradient paths 139
5.8   Simulated annealing paths 144
5.9   Simulated annealing sequence with two modes 146
5.10 Simulated annealing sequence for four schedules 147
5.11 Monte Carlo approximations of a probit marginal 149
5.12 EM sequences for a normal censored likelihood 155
5.13 Multiple-mode EM 156
5.14 MCEM on logit model 162

6.1   Metropolis output from a beta target 173
6.2   Metropolis simulations from a beta target 174
6.3   Output of a gamma accept–reject algorithm 177
6.4   Metropolis–Hastings schemes for a Cauchy target 179
6.5   Cumulative coverage for a Cauchy target 180
6.6   Fitting the braking data 181
6.7   Random walk proposals with different scales 184
6.8   Scale impact on mixture exploration 185
6.9   Langevin samples for probit posterior 187
6.10 Langevin samples for mixture posterior 189
6.11 Cumulative mean plots with different scales 193

7.1   Histograms from the Gibbs sampler of Example 7.2 203
7.2   Histograms from the Gibbs sampler of Example 7.3 205
7.3   Histograms from the Gibbs sampler of Example 7.5 208
7.4   Histograms of the posterior distributions from Example 7.6 211
7.5   Histograms from the Gibbs sampler of Example 7.7 212
7.6   Histograms from the Gibbs sampler of Example 7.8 214
7.7   Gibbs output for mixture posterior 217
7.8   Simple slice sampler 219

7.9   Logistic slice sampler 221
7.10 Histograms from the pump failure data of Example 7.12 223
7.11 Autocorrelations from a Gibbs sampler 225
7.12 Autocovariance plots for the Gibbs sampler of model (7.7) 226
7.13 Histograms of λ in Example 7.16 229
7.14 Histograms from the Gibbs sampler of (7.12) 233

8.1   Raw coda output for the random effect logit model 244
8.2   Empirical cdfs for the random effect logit parameters 244
8.3   Plot of successive Kolmogorov–Smirnov statistics 246
8.4   Comparison of two MCMC scales for the noisy AR model 251
8.5   Multiple MCMC runs for the noisy AR model 252
8.6   Gelman and Rubin's evaluation for the noisy AR model 254
8.7   Gelman and Rubin's evaluation for the pump failure model 255
8.8   Fixed-width batch sampling variance estimation for the pump
      failure model 259
8.9   Degenerating MCMC adaptation for the pump failure model ... 261
8.10 Nonconverging non-parametric MCMC adaptation for the
      noisy AR model 262
8.11 Mode-recovering non-parametric MCMC adaptation for the
      noisy AR model 263

List of Examples

1.1   Bootstrapping simple linear regression 25

2.1   Exponential variable generation 44
2.2   Transformations of exponentials 46
2.3   Normal variable generation 47
2.4   Discrete random variables 48
2.5   Poisson random variables 49
2.6   Negative binomial random variables as mixtures 50
2.7   Accept–reject for beta variables 53
2.8   Continuation of Example 2.7 55

3.1   Precision of integrate 62
3.2   integrate versus area 63
3.3   Monte Carlo convergence 65
3.4   Precision of a normal cdf approximation 67
3.5   Tail probability approximation 70
3.6   Beta posterior importance approximation 71
3.7   Continuation of Example 3.6 77
3.8   Importance sampling with infinite variance 79
3.9   Selection of the importance sampling function 82
3.10   Probit posterior importance sampling approximation 83

4.1   Monitoring with the CLT 91
4.2   Cauchy prior 92
4.3   Continuation of Example 4.2 96
4.4   Continuation of Example 4.3 100
4.5   Continuation of Example 4.4 105
4.6   Student's t expectation 107
4.7   James–Stein estimation 112
4.8   Continuation of Example 4.1 114
4.9   Cauchy posterior with antithetic variables 115

4.10 Continuation of Example 4.9 118
4.11 Logistic regression 119

5.1   Maximizing a Cauchy likelihood 127
5.2   Mixture model likelihood 128
5.3   A first Monte Carlo maximization 131
5.4   Continuation of Example 5.1 131
5.5   Continuation of Example 5.4 133
5.6   Minimization of a complex function 135
5.7   Continuation of Example 5.6 138
5.8   Continuation of Example 5.3 142
5.9   Simulated annealing for a normal mixture 144
5.10 Continuation of Example 5.6 145
5.11 Bayesian analysis of a simple probit model 147
5.12 Missing-data mixture model 150
5.13 Censored–data likelihood 151
5.14 Continuation of Example 5.13 153
5.15 EM for a normal mixture 154
5.16 Missing–data multinomial model 158
5.17 Random effect logit model 159

6.1   Metropolis–Hastings algorithm for beta variables 172
6.2   Cauchys from normals 177
6.3   Metropolis–Hastings for regression 179
6.4   Normals from uniforms 183
6.5   Metropolis–Hastings for mixtures 183
6.6   Probit regression 186
6.7   Continuation of Example 6.5 187
6.8   Model selection 188
6.9   Acceptance rates: normals from double exponentials 192
6.10 Continuation of Example 6.4 195

7.1   Normal bivariate Gibbs 201
7.2   Generating beta-binomial random variables 202
7.3   Fitting a one-way hierarchical model 202
7.4   Normal multivariate Gibbs 206
7.5   Extension of Example 7.3 207
7.6   Censored-data Gibbs 210
7.7   Grouped multinomial data 211
7.8   More grouped multinomial data 213
7.9   Gibbs for normal mixtures 215
7.10 A first slice sampler 218
7.11 Logistic regression with the slice sampler 219
7.12 A Poisson hierarchy 222
7.13 Correlation in a bivariate normal 224

7.14 Continuation of Example 7.5 225
7.15 Normal bivariate Gibbs revisited 227
7.16 Poisson counts with missing data 228
7.17 Metropolis within Gibbs illustration 230
7.18 Conditional exponential distributions—-nonconvergence 232
7.19 Improper random effects posterior 233

8.1   Random effect logit model 242
8.2   Poisson hierarchical model 245
8.3   Metropolis–Hastings random walk on AR(1) model 249
8.4   Continuation of Example 8.3 250
8.5   Continuation of Example 8.2 254
8.6   Continuation of Example 8.5 258
8.7   Another approach to Example 8.2 259
8.8   Continuation of Example 8.4 261
8.9   Adaptive MCMC for anova 264
8.10 Continuation of Example 8.7 265
8.11 Continuation of Example 8.8 266

1

Basic R Programming

"You're missing the big picture," he told her. "A good album should be more than the sum of its parts."

Ian Rankin
Exit Music

Reader's guide

The **Reader's guide** is a section that will start each chapter by providing comments on its contents. It also usually contains indications of the purpose of the chapter and its links with other chapters.

This first chapter is where we introduce the programming language R, which we use to implement and illustrate our algorithms. We discuss here input and output, data structures, and basic programming commands for this language. It is thus a crucial chapter for these new to R, but it will unavoidably feel unsatisfactory because the coverage of those notions will most likely be too sketchy for most readers. For those already familiar with R, or simply previously exposed to another introduction to R, this chapter will undoubtedly feel mostly familiar and act as a refresher, maybe prodding them to delve deeper into the R language using a reference book. The similarity with the introductory chapter of *Bayesian Core* is not coincidental, as we used the same skeleton as in Marin and Robert (2007).

C.P. Robert, G. Casella, *Introducing Monte Carlo Methods with R*, Use R,
DOI 10.1007/978-1-4419-1576-4_1, © Springer Science+Business Media, LLC 2010

1.1 Introduction

This chapter only attempts at introducing R to newcomers in a few pages and, as such, it should not be considered as a proper introduction to R. Entire volumes, such as the monumental *R Book* by Crawley (2007), the introduction by Dalgaard (2002), and the focused R data manipulation by Spector (2009), are dedicated to the practice of this language, and therefore additional efforts (besides reading this chapter) will be required from the reader to sufficiently master the language.[1] However, before discouraging anyone, let us comfort you with the fact that:

a. The syntax of R is simple and logical enough to quickly allow for a basic understanding of simple R programs, as should become obvious in a few paragraphs.
b. The best, and in a sense the only, way to learn R is through trial-and-error on simple and then more complex examples. Reading the book with a computer available nearby is therefore the best way of implementing this recommendation.

In particular, the embedded help commands `help()` and `help.search()` are very good starting points to gather information about a specific function or a general issue, even though more detailed manuals are available both locally and on-line. Note that `help.start()` opens a Web browser linked to the local manual pages.

One may first wonder why we support using R as the programming interface for this introduction to Monte Carlo methods, since there exist other languages, most (all?) of them faster than R, like Matlab, and even free, like C or Python. We obviously have no partisan or commercial involvement in this language. Rather, besides the ease of presentation, our main reason for this choice is that the language combines a sufficiently high power (for an interpreted language) with a very clear syntax both for statistical computation and graphics. R is a flexible language that is *object-oriented* and thus allows the manipulation of complex data structures in a condensed and efficient manner. Its graphical abilities are also remarkable, with possible interfacing with a text processor such as LATEX with the package Sweave. R offers the additional advantages of being a free and open-source system under the GNU General Public Licence principle, with constant upgrades and improvements from the statistics community,[2] as well as numerous (free) Web-based tutorials and user's manuals,[3] and running on all platforms, including both Apple's Mac

---

[1] If you decide to skip this chapter, be sure to at least print the handy R Reference Card available at http://cran.r-project.org/doc/contrib/Short-refcard.pdf that summarizes, in four pages, the major commands of R.

[2] There is even an R newsletter, *R-News*, which is available on cran.r-project.org/doc/Rnews.

[3] This means it is unlikely that you will need to acquire an R programming book in order to get a proper understanding of the R language, even though it may

and Microsoft Windows (and, obviously, under the Linux and Unix operating systems). R provides a powerful *interface* that can integrate programs written in other languages such as C, C++, Fortran, Perl, Python, and Java. Not only can you keep programming in your usual language, but integrating programs written by others in an alien language then becomes (mostly) seamless, as seen in Chapter 8 with the package amcmc. At last, it is increasingly common to see people who develop new methodology simultaneously producing an R package in support of their approach and to back up introductory statistics courses with illustrations in R, as shown by the expanding Use R! Springer series in which this book is published.

One choice we have *not* addressed above is "why R and not BUGS?" BUGS (which stands for Bayesian inference using Gibbs sampling) is a Bayesian analysis software developed since the early 1990's, mostly by researchers from the Medical Research Council (MRC) at Cambridge University. The most common version is WinBugs, working under Windows, but there also exists an open-source version called OpenBugs. So, to return to the initial question, we are not addressing the possible links and advantages of BUGS simply because the purpose is different. Our goal is to give a deep but intuitive understanding of simulation tools for statistical inference, not (directly) help in the construction of Bayesian models and the Bayesian estimation of their parameters. While access to Monte Carlo specifications is possible in BUGS, most computing operations are handled by the software itself, with the possible outcome that the user does not bother about this side of the problem and instead concentrates on Bayesian modeling. Thus, while R can be easily linked with BUGS and simulation can be done via BUGS, we think that a lower-level language such as R is more effective in bringing you in touch with simulation imperatives. Note that this perspective would not hold so tightly for a book on computational statistics, as Albert (2009).

1.2 Getting started

The R language is straightforward to install: It can be downloaded (obviously free) from one of the numerous CRAN (Comprehensive R Archive Network) mirror Websites around the world.[4] (Note that it is resident in most current Linux kernels.)

At this stage, we do not cover installation of the R package and thus assume that (a) R is installed on the machine you want to work with and (b) that you have managed to launch it (in most cases, you simply have to click on

---

prove useful at a later stage. See http://cran.r-project.org/manuals.html for manuals available on-line.

[4] The main CRAN Website is http://cran.r-project.org/.

the proper icon). You should then obtain a terminal window whose first lines resemble the following, most likely with a more recent version:

```
R version 2.5.1 (2007-06-27)
Copyright (C) 2007 The R Foundation for Statistical Computing
ISBN 3-900051-07-0

R is free software and comes with ABSOLUTELY NO WARRANTY.
You are welcome to redistribute it under certain conditions.
Type 'license()' or 'licence()' for distribution details.

R is a collaborative project with many contributors.
Type 'contributors()' for more information and
'citation()' on how to cite R or R packages in publications.

Type 'demo()' for some demos, 'help()' for on-line help, or
'help.start()' for an HTML browser interface to help.
Type 'q()' to quit R.

>
```

Neither this austere beginning nor the prospect of using a line editor should put you off, though, as there are many other ways of inputting and outputting commands and data, as we shall soon see! The final line above with the symbol > is not a typo but rather means that the R software is waiting for a command from the user (i.e., you). This character > at the beginning of each line in the executable window is called the *prompt* and precedes the line command, which is terminated by pressing the RETURN key. At this early stage, all commands will be passed as line commands, and you should thus spot commands thanks to this symbol.

↯ Make sure to remember that exiting R can be done by typing q() after the prompt, as in

```
> q()
Save workspace image? [y/n/c]: n
```

the options being proposed in the line right after the prompt having to do with the storage of the command history and the produced objects, something you can ignore at this stage. Commands and programs that need to be stopped during their execution, for instance because they take too long or too much memory to complete or because they involve a programming mistake such as an infinite loop, can be stopped by the Control-C double-key action without exiting the R session.

> **Exercise 1.1** To start with a limited challenge, using the final lines above the prompt, you could embark on an on-line visit of the main features of R by typing demo() after the prompt (make sure to test demo(image) and demo(graphics) to get an idea of the great graphical abilities of R). More statistical demos are also available, as listed by demo().

For memory and efficiency reasons, R does not install all the available functions and programs when launched but only the basic *packages* that it requires to run properly. Those basic packages are base, stats, graphics, nmle, and lattice. Additional packages can be loaded via the library command, as in

```
> library(combinat) # combinatorics utilities
> library(datasets) # The R Datasets Package
```

and the entire list of available packages is provided by library(). (The symbol # in the prompt lines above indicates a comment: All characters following # until the end of the command line are ignored. Comments are recommended to improve the readability of your programs.) There exist hundreds of packages available on the Web.[5] Installing a new package such as the package mcsm that is associated with this book is done by downloading the file from the Web depository and calling

```
> install.package("mcsm")
```

or

```
> download.package("mcsm")
```

For a given package, the install command obviously needs to be executed only once, while the library call is required each time R is launched (as the corresponding package is not kept as part of the .RData file, whose role is explained below in Section 1.9). Thus, it is good practice to include calls to required libraries within your R programs in order to avoid error messages when launching them.

1.3 R objects

As with many advanced programming languages, R distinguishes between several types of *objects*. Those types include scalar, vector, matrix, time series, data frames, functions, or graphics. An R object is mostly characterized by a *mode* that describes its contents and a *class* that describes its structure. The R function str applied to any R object, including R functions, will show its structure. The different modes are

---

[5] Packages that have been validated and tested by the R core team are listed at http://cran.r-project.org/src/contrib/PACKAGES.html.

- null (empty object),
- logical (TRUE or FALSE),
- numeric (such as 3, 0.14159, or 2+sqrt(3)),
- complex, (such as 3-2i or complex(1,4,-2)), and
- character (such as "Blue", "binomial", "male", or "y=a+bx"),

and the main classes are vector, matrix, array, factor, time-series, data.frame, and list. Heterogeneous objects such as those of the list class can include elements with various modes. Manual entries about those classes can be obtained via the help commands help(data.frame) or ?matrix for instance.

R can operate on most of those types as a regular function would operate on a scalar, which is a feature that should be exploited as much as possible for compact and efficient programming. The fact that R is interpreted rather than compiled involves many subtle differences, but a major issue is that all variables in the system are evaluated and stored at every step of R programs. This means that loops in R are enormously time-consuming and should be avoided at all costs! Therefore, using the shortcuts offered by R in the manipulation of vectors, matrices, and other structures is a must.

1.3.1 The vector class

As indicated logically by its name, the vector object corresponds to a mathematical vector of elements of the same type, such as (TRUE,TRUE,FALSE,TRUE) or (1,2,3,5,7,11). Creating small vectors can be done using the R command c() as in

```
> a=c(2,6,-4,9,18)
```

This fundamental function combines or concatenates terms together. For instance,

```
> d=c(a,b)
```

concatenates the two vectors a and b into a new vector d. Note that decimal numbers should be encoded with a dot, character strings in quotes " ", and logical values with the character strings TRUE and FALSE or with their respective abbreviations T and F. Missing values are encoded with the character string NA. The option recursive=TRUE in c() allows breaking down a list into its individual components.

⚡ Being able to use abbreviations in R is quite handy, but this may lead to confusion! In particular, the use of T instead of TRUE is only valid if T is not defined otherwise in the current R session. Since T is a standard symbol in Monte Carlo simulation, often denoting the number of iterations, this may create unsuspected problems. For instance, using

```
> sort(weit,dec=T)
Error in sort(weit/sum(weit), dec = T) :
        'decreasing' must be a length-1 logical vector.
Did you intend to set 'partial'?
```

resulted in an error message because T was already defined in the R program as 10^3.

In Figure 1.1, we give a few illustrations of the use of vectors in R. The character + indicates that the console is waiting for a supplementary instruction, which is useful when typing long expressions. The assignment operator is =, not to be confused with ==, which is the Boolean operator for equality. An older assignment operator is <-, as in

```
> x <- c(3,6,9)
```

and, for compatibility reasons, it still remains functional, but it should be ignored to ensure cleaner programming. (As pointed out by Spector (2009), an exception is when using system.time, briefly described in Figure 1.13, since = is then used to identify keywords, although = can preserve its initial purpose if curly brackets { and } delimit the allocation commands.)

↯ A misleading feature of the assignment operator <- is found in Boolean expressions such as

```
> if (x[1]<-2) ...
```

which is supposed to test whether or not x[1] is less than -2 but ends up allocating 2 to x[1], erasing its current value! Note also that using

```
> if (x[1]=-2) ...
```

mistakenly instead of (x[1]==-2) has the same consequence.

Exercise 1.2 Propose a valid R expression to overcome the difficulty with the erroneous Boolean expression if (x[1]<-2) found above. (*Hint:* Adding a space in the expression is sufficient.)

If you hit RETURN after the command if (x[1]<-2), you will see a + appear on the next line instead of the prompt >. This means that your R command is not complete and that the system is waiting for you to complete it. This option is quite handy when handling long commands.

Note that new R objects are simply defined by assigning them a value, as in the first line of Figure 1.1, without a preliminary declaration of type (as in the C language).

```
> a=c(5,5.6,1,4,-5)        build the object a containing a numeric vector
                           of dimension 5 with elements 5, 5.6, 1, 4, −5
> a[1]                     display the first element of a
> b=a[2:4]                 build the numeric vector b of dimension 3
                           with elements 5.6, 1, 4
> d=a[c(1,3,5)]            build the numeric vector d of dimension 3
                           with elements 5, 1, −5
> 2*a                      multiply each element of a by 2
                           and display the result
> b%%3                     provides each element of b modulo 3
> d%/%2.4                  computes the integer division of each element of d by 2.4
> e=3/d                    build the numeric vector e of dimension 3
                           and elements 3/5, 3, −3/5
> log(d*e)                 multiply the vectors d and e term by term
                           and transform each term into its natural logarithm
> sum(d)                   calculate the sum of d
> length(d)                display the length of d
> t(d)                     transpose d, the result is a row vector
> t(d)%*%e                 scalar product between the row vector t(b) and
                           the column vector e with identical length
> t(d)*e                   elementwise product between two vectors
                           with identical lengths
> g=c(sqrt(2),log(10))     build the numeric vector g of dimension 2
                           and elements $\sqrt{2}$, log(10)
> e[d==5]                  build the subvector of e that contains the
                           components e[i] such that d[i]=5
> a[-3]                    create the subvector of a that contains
                           all components of a but the third.[6]
> is.vector(d)             display the logical expression TRUE if
                           a vector and FALSE else
```

Fig. 1.1. Illustrations of the processing of vectors in R.

Note the convenient use of Boolean expressions to extract subvectors from a vector without having to resort to a component-by-component test (and hence a loop). The quantity d==5 is itself a vector of Booleans, while the number of components satisfying the constraint can be computed by sum(d==5). The ability to apply scalar functions to vectors as a whole is also a major advantage of R. In the event the function depends on a parameter or an option, this quantity can be entered as in

```
> e=gamma(e^2,log=T)
```

which returns the vector with components $\log \Gamma(e_i^2)$. Functions that are specially designed for vectors include, for instance, sample, permn, order, sort, and rank, which all have to do with manipulating the order in which the components of the vector occur. (Note that permn is part of the combinat

---

[6] Positive and negative indices cannot be used simultaneously.

library, as indicated when typing `help.search("permn")`, which returns a `permn(combinat)` among its matching entries.)

Exercise 1.3 Test the `help` command on the functions seq, sample, and order. (*Hint:* Start with `help(help)`.)

Exercise 1.4 Explain the difference between the functions order and rank. For the function rep., explain the difference between the options times, length.out, and each.

Besides their numeric and logical indexes, the components of a vector can also be identified by names. For a given vector x, `names(x)` is a vector of characters of the same length as x. This additional attribute is most useful when dealing with real data, where the components have a meaning such as "unemployed" or "democrat". Those names can also be erased by the command

```
> names(x)=NULL
```

The : operator found in Figure 1.1 is a very useful device that defines a consecutive sequence, but it is also fragile in that reverse sequences do not always produce what is expected.[7] For one thing, `1:n-1` is interpreted as `(1:n)-1` rather than `1:(n-1)`. For another, while `3:1` returns the vector `c(3,2,1)`, the command `1:0` returns `c(1,0)`, which may or may not be okay depending on the circumstances. For instance, `a[1:0]` will only return `a[1]`, and this may not be the limiting case the programmer had in mind. Note also that `a[0]` does not produce an error message but a vector with length zero.

Exercise 1.5 Show that the alternative `seq(1,n-1,by=1)` does not suffer from the same drawbacks as `1:(n-1)`. Find a modification of by=1 that handles the case where $n \leq 1$.

1.3.2 The matrix, array, and factor classes

The matrix class provides the R representation of matrices. A typical entry is, for instance,

```
> x=matrix(vec,nrow=n,ncol=p)
```

which creates an $n \times p$ matrix whose elements are those of the vector vec, assuming this vector is of dimension np. An important feature of this entry is that, in a somewhat unusual way, the components of vec are stored by column, which means that `x[1,1]` is equal to `vec[1]`, `x[2,1]` is equal to

---

[7] This difficulty was pointed out by Radford Neal.

vec[2], and so on, except if the option byrow=T is used in matrix. (Because
of this choice of storage convention, working on R matrices columnwise is
faster then working rowwise.) Note also that, if vec is of dimension $n \times p$, it is
not necessary to specify both the nrow=n and ncol=p options in matrix. One
of those two parameters is sufficient to define the matrix. On the other hand,
if vec is *not* of dimension $n \times p$, matrix(vec,nrow=n,ncol=p) will create an
$n \times p$ matrix with the components of vec repeated the appropriate number of
times. For instance,

```
> matrix(1:4,ncol=3)
      [,1] [,2] [,3]
[1,]    1    3    1
[2,]    2    4    2
Warning message:
data length [4] is not a submultiple or multiple of the number
of columns [3] in matrix in: matrix(1:4, ncol = 3)
```

produces a 2×3 matrix along with a warning message that something may
be missing in the call to matrix. Note again that $1, 2, 3, 4$ are entered consecutively when following the column (or *lexicographic*) order. Names can be
given to the rows and columns of a matrix using the rownames and colnames
functions.

In some situations, it is useful to remember that an R matrix can also be
used as a vector. If x is an $n \times p$ matrix, x[i,j]=x[i+n*(j-1)] is equal
to x[i,j], i.e., x can also be manipulated as a vector made of the columns
of vec piled on top of one another. For instance, x[x>5] is a vector, while
x[x>5]=0 modifies the right entries in the matrix x. Conversely, vectors can
be turned into $p \times 1$ matrices by the command as.matrix. Note that x[1,]
produces the first row of x as a vector rather than as a $p \times 1$ matrix.

R allows for a wide range of manipulations on matrices, both termwise and
in the classical matrix algebra perspective. For instance, the standard matrix
product is denoted by %*%, while * represents the term-by-term product. (Note
that taking the product a%*%b when the number of columns of a differs from
the number of rows of b produces an error message.) Figure 1.2 gives a few
examples of matrix-related commands. The apply function is particularly easy
to use for functions operating on matrices by row or column.

The function diag can be used to extract the vector of the diagonal elements of a matrix, as in diag(a), or to create a diagonal matrix with a
given diagonal, as in diag(1:10). Since matrix algebra is central to good
programming in R, as matrix programming allows for the elimination of time-
consuming loops, it is important to be familiar with matrix manipulation. For
instance, the function crossprod replaces the product t(x)%*%y on either
vectors or matrices by crossprod(x,y) more efficiently:

```
> system.time(crossprod(1:10^6,1:10^6))
```

```
> x1=matrix(1:20,nrow=5)              build the numeric matrix x1 of dimension
                                      5 × 4 with first row 1, 6, 11, 16
> x2=matrix(1:20,nrow=5,byrow=T)      build the numeric matrix x2 of dimension
                                      5 × 4 with first row 1, 2, 3, 4
> a=x3%*%x2                           matrix summation of x2 and x3
> x3=t(x2)                            transpose the matrix x2
> b=x3%*%x2                           matrix product between x2 and x3,
                                      with a check of the dimension compatibility
> c=x1*x2                             term-by-term product between x1 and x2
> dim(x1)                             display the dimensions of x1
> b[,2]                               select the second column of b
> b[c(3,4),]                          select the third and fourth rows of b
> b[-2,]                              delete the second row of b
> rbind(x1,x2)                        vertical merging of x1 and x2
> cbind(x1,x2)                        horizontal merging of x1 and x2
> apply(x1,1,sum)                     calculate the sum of each row of x1
> as.matrix(1:10)                     turn the vector 1:10 into a 10 × 1 matrix
```

Fig. 1.2. Illustrations of the processing of matrices in R.

```
    user   system  elapsed
   0.016    0.048    0.066
> system.time(t(1:10^6)%*%(1:10^6))
    user   system  elapsed
   0.084    0.036    0.121
```

(You can also check the symmetric function tcrossprod.)

Eigenanalysis of square matrices is also included in the base package. For instance, chol(m) returns the upper triangular factor of the Choleski decomposition of m; that is, the matrix R such that $R^T R$ is equal to m. Similarly, eigen(m) returns a list (see Section 1.3.3) that contains the eigenvalues of m (some of which can be complex numbers) as well as the corresponding eigenvectors (some of which are complex if there are complex eigenvalues). Related functions are svd and qr, which provide the singular values and the QR decomposition of their argument, respectively. Note that the inverse M^{-1} of a matrix M can be found either by solve(M) (recommended) or ginv(M), which requires downloading the library MASS and also produces generalized inverses (which may be a mixed blessing since the fact that a matrix is not invertible is not signaled by ginv). Special versions of solve are backsolve and forwardsolve, which are restricted to upper and lower diagonal triangular systems, respectively. Note also the alternative of using chol2inv which returns the inverse of a matrix m when provided by the Choleski decomposition chol(m).

Structures with more than two indices are represented by *arrays* and can also be processed by R commands, for instance x=array(1:50,c(2,5,5)), which gives a three-entry table of 50 terms. Once again, they can also be interpreted as vectors.

The `apply` function used in Figure 1.2 is a very powerful device that operates on arrays and, in particular, matrices. Since it can return arrays, it bypasses calls to multiple loops and makes for (sometimes) quicker and (always) cleaner programs. It should not be considered as a panacea, however, as `apply` hides calls to loops inside a single command. For instance, a comparison of `apply(A, 1, mean)` with `rowMeans(A)` shows the second version is about 200 times faster. Using linear algebra whenever possible is therefore a more efficient solution. Spector (2009, Section 8.7) gives a detailed analysis of the limitations of `apply` and the advantages of vectorization in R.

A `factor` is a vector of characters or integers used to specify a discrete classification of the components of other vectors with the same length. Its main difference from a standard vector is that it comes with a level attribute used to specify the possible values of the factor. This structure is therefore appropriate to represent qualitative variables. R provides both ordered and unordered factors, whose major appeal lies within model formulas, as illustrated in Figure 1.3. Note the subtle difference between `apply` and `tapply`.

```
> state=c("tas","tas","sa","sa","wa")   create a vector with five values
> statef=factor(state)                  distinguish entries by group
> levels(statef)                        give the groups
> incomes=c(60,59,40,42,23)             create a vector of incomes
> tapply(incomes,statef,mean)           average the incomes for each group
> statef=factor(state,                  define a new level with one more
+ levels=c("tas","sa","wa","yo"))       group than observed
> table(statef)                         return statistics for all levels
```

Fig. 1.3. Illustrations of the factor class.

1.3.3 The list and data.frame classes

A list in R is a rather loose object made of a collection of other arbitrary objects known as its *components*.[8] For instance, a list can be derived from n existing objects using the function list:

    a=list(name_1=object_1,...,name_n=object_n)

This command creates a list with n arguments using `object_1,...,object_n` for the components, each being associated with the argument's name, `name_i`. For instance, `a$name_1` will be equal to `object_1`. (It can also be represented as `a[[1]]`, but this is less practical, as it requires some bookkeeping of the order of the objects contained in the list.) Lists are very useful in preserving information about the values of variables used within R functions in the sense

---
[8] Lists can contain lists as elements.

that all relevant values can be put within a list that is the output of the corresponding function (see Section 1.7 for details about the construction of functions in R). Most standard functions in R, for instance eigen in Figure 1.4, return a list as their output. Note the use of the abbreviations vec and val in the last line of Figure 1.4. Such abbreviations are acceptable as long as they do not induce confusion. (Using res$v would not work!)

```
> li=list(num=1:5,y="color",a=T)            create a list with three arguments
> a=matrix(c(6,2,0,2,6,0,0,0,36),nrow=3)    create a (3,3) matrix
> res=eigen(a,symmetric=T)                   diagonalize a and
> names(res)                                 produce a list with two
                                             arguments: vectors and values
> res$vectors                                vectors arguments of res
> diag(res$values)                           create the diagonal matrix
                                             of eigenvalues
> res$vec%*%diag(res$val)%*%t(res$vec)       recover a
```

Fig. 1.4. Chosen features of the list class.

The local version of apply is lapply, which computes a function for each argument of the list

```
> x = list(a = 1:10, beta = exp(-3:3),
+ logic = c(TRUE,FALSE,FALSE,TRUE))
> lapply(x,mean) #compute the empirical means
$a
[1] 5.5
$beta
[1] 4.535125
$logic
[1] 0.5
```

provided each argument is of a mode that is compatible with the function argument (i.e., is numeric in this case). A "user-friendly" version of lapply is sapply, as in

```
> sapply(x,mean)
       a      beta     logic
5.500000 4.535125 0.500000
```

The last class we briefly mention here is the data frame. A data frame is a list whose elements are possibly made of differing modes and attributes but have the same length, as in the example provided in Figure 1.5. A data frame can be displayed in matrix form, and its rows and columns can be extracted using matrix indexing conventions. A list whose components satisfy the restrictions imposed on a data frame can be coerced into a data frame using the function `as.data.frame`. The main purpose of this object is to import data from an external file by using the `read.table` function.

> v1=sample(1:12,30,rep=T)	simulate 30 independent uniform random variables on $\{1, 2, \ldots, 12\}$
> v2=sample(LETTERS[1:10],30,rep=T)	simulate 30 independent uniform random variables on $\{a, b, \ldots, j\}$
> v3=runif(30)	simulate 30 independent uniform random variables on $[0, 1]$
> v4=rnorm(30)	simulate 30 independent realizations from a standard normal distribution
> xx=data.frame(v1,v2,v3,v4)	create a data frame

Fig. 1.5. Definition of a data frame.

1.4 Probability distributions in R

R is primarily a statistical language. It is therefore well-equipped with probability distributions. As described in Table 1.1, all standard distributions are available, with a clever programming shortcut: A "core" name, such as `norm`, is associated with each distribution, and the four basic associated functions, namely the cdf, the pdf, the quantile function, and the simulation procedure, are defined by appending the prefixes `d`, `p`, `q`, and `r` to the core name, such as `dnorm`, `pnorm`, `qnorm`, and `rnorm`. Obviously, each function requires additional entries, as in `pnorm(1.96)` or `rnorm(10,mean=3,sd=3)`. Recall that `pnorm` and `qnorm` are inverses of one another.

Exercise 1.6 Study the properties of the R function `lm` using simulated data as in

```
> x=rnorm(20)
> y=3*x+5+rnorm(20,sd=0.3)
> reslm=lm(y~x)
> summary(reslm)
```

The simulation aspects related to the normal distribution (and to these other standard distributions) will be discussed in detail in Chapter 2.

1.5 Basic and not-so-basic statistics

R is designed by statisticians, as the logical continuation of the former S-plus language, and, as such, it offers a very wide range of statistical packages that cover the entire spectrum of statistics. The battery of these (classical) statistical tools, ranging from descriptive statistics to non-parametric density

Table 1.1. Standard distributions with R core name.

Distribution	Core	Parameters	Default Values
Beta	beta	shape1, shape2	
Binomial	binom	size, prob	
Cauchy	cauchy	location, scale	0, 1
Chi-square	chisq	df	
Exponential	exp	1/mean	1
F	f	df1, df2	
Gamma	gamma	shape,1/scale	NA, 1
Geometric	geom	prob	
Hypergeometric	hyper	m, n, k	
Log-normal	lnorm	mean, sd	0, 1
Logistic	logis	location, scale	0, 1
Normal	norm	mean, sd	0, 1
Poisson	pois	lambda	
Student	t	df	
Uniform	unif	min, max	0, 1
Weibull	weibull	shape	

estimation and generalized linear models, cannot be provided in this book, but we refer you to, for instance, Dalgaard (2002) or Venables and Ripley (1999) for a detailed introduction.

At the most basic level, descriptive statistics can be obtained for any object with numerical entries. For instance, mean (possibly trimmed), var, sd, median, quantile, and summary produce standard estimates for the samples on which they are called. Note that, due to the choice of an unbiased version of this estimator that involves dividing the sum of squares by $n-1$ instead of dividing by n, the variance estimator of a single observation is NA rather than 0.

When applied to a matrix x, the output of var(x) differs from the output of sd(x)^2

```
> b=matrix(1:9,ncol=3)
> var(b)
      [,1] [,2] [,3]
[1,]    1    1    1
[2,]    1    1    1
[3,]    1    1    1
> sd(b)^2
[1] 1 1 1
```

because the former returns an estimate of the covariance between the columns of x, while the latter produces an estimate of the variances of the columns. Note that the definition of b only specifies the number of columns, 3 in this case, and thus assumes that the length of the vector is a multiple of 3. (If it is not, R produces a warning that the data length is not a submultiple or multiple of the number of columns.)

Classical hypothesis tests, such as the equality of two means or the equality of two variances, can be conducted using standard functions. Typing `help.search("test")` will produce a list of tests that most likely contains more tests than you have previously heard of. For example, checking that the mean of a normal sample with unknown variance is zero can be conducted using the t test (Casella and Berger, 2001) as

```
> x=rnorm(25) #produces a N(0,1) sample of size 25
> t.test(x)
```

```
        One Sample t-test

data:  x
t = -0.8168, df = 24, p-value = 0.4220
alternative hypothesis: true mean is not equal to 0
95 percent confidence interval:
 -0.4915103  0.2127705
sample estimates:
 mean of x
-0.1393699
```

whose outcome can be interpreted as providing a p-value of 0.4220 (i.e., a fairly large probability of observing a larger empirical average $\bar{x}$ than the one just observed, -0.139) and hence as concluding that the data do not contradict the null hypothesis.

As pointed out previously, all but the most basic R functions return lists as their output (or value). For instance, when running `t.test` above, the output involves nine arguments:

```
> out=t.test(x)
> names(out)
[1] "statistic" "parameter"  "p.value" "conf.int" "estimate"
[6] "null.value" "alternative" "method"   "data.name"
```

which can be handled separately, as for instance in `as.numeric(out$est)^2`.

Similarly, the presence of correlation between two variables can be tested by cor.test, as in the example

```
> attach(faithful) #resident dataset
> cor.test(faithful[,1],faithful[,2])
```

```
            Pearson's product-moment correlation

data:  faithful[, 1] and faithful[, 2]
t = 34.089, df = 270, p-value < 2.2e-16
alternative hypothesis: true correlation is not equal to 0
95 percent confidence interval:
 0.8756964 0.9210652
sample estimates:
     cor
0.9008112
```

which concludes that the data, faithful, made of eruptions and waiting, which correspond to the eruption times and the waiting times of the Old Faithful geyser in Yellowstone National Park, has its two variables most certainly correlated.

Non-parametric tests such as the one-sample and two-sample Kolmogorov–Smirnov adequation tests (ks.test), Shapiro's normality test (shapiro.test), Kruskall–Wallis homogeneity test (kruskal.test), and Wilcoxon rank tests (wilcox.test) are available. For instance, testing for normality on the faithful dataset leads to

```
> ks.test(jitter(faithful[,1]),pnorm)

            One-sample Kolmogorov-Smirnov test

data:  jitter(faithful[, 1])
D = 0.9486, p-value < 2.2e-16
alternative hypothesis: two-sided

> shapiro.test(faithful[,2])

            Shapiro-Wilk normality test

data:  faithful[, 2]
W = 0.9221, p-value = 1.016e-10

> wilcox.test(faithful[,1])

            Wilcoxon signed rank test with continuity correction

data:  faithful[, 1]
V = 37128, p-value < 2.2e-16
alternative hypothesis: true location is not equal to 0
```

In the first command line above, the function jitter is used to perturb each entry in the dataset in order to remove the ties within it. Otherwise, the p-value cannot be computed:

```
Warning message:
cannot compute correct p-values with ties in:
ks.test(faithful[, 1], pnorm)
```

This function is also quite handy when plotting datasets with ties.

> Most R functions require arguments, and most of them have default values for at least some arguments. For instance, the Wilcoxon test `wilcox.test` has `mu=0` as its default location to be tested. Those default values are indicated on the help page of the functions.

Non-parametric kernel density estimates can similarly be constructed via the function density and are quite amenable to calibration, from the choice of the kernel to the choice of the bandwidth (see Venables and Ripley, 1999). In our case, they will be handy as sample-based proposals when designing MCMC algorithms in Chapters 6 and 8. Spline modeling also is available via the functions spline and splinefun. Non-parametric regression can be performed via the loess function or using *natural splines*.

For instance, the data constructed as

```
> Nit = c(0,0,0,1,1,1,2,2,2,3,3,3,3,4,4,4,6,6,6)
> AOB =c(4.26,4.15,4.68,6.08,5.87,6.92,6.87,6.25,
+ 6.84,6.34,6.56,6.52,7.39,7.38,7.74,7.76,8.14,7.22)
```

reports on the relationship between nitrogen level in soil (coded 0,1,2,3,4,6) and abundance of a bacteria called AOB, reproduced on the left-hand side of Figure 1.6. The loess and natural spline fits are obtained via the R code

```
> AOBm=tapply(AOB,Nit,mean)                  #means of AOB
> Nitm=tapply(Nit,Nit,mean)                  #means of Nit
> plot(Nit,AOB,xlim=c(0,6),ylim=c(min(AOB),max(AOB)),pch=19)
> fitAOB=lm(AOBm~ns(Nitm,df=2))              #natural spline
> xmin=min(Nit);xmax=max(Nit)
> lines(seq(xmin,xmax,.5),                   #fit to means
+     predict(fitAOB,data.frame(Nitm=seq(xmin,xmax,.5))))
> fitAOB2=loess(AOBm~Nitm,span = 1.25)   #loess
> lines(seq(xmin,xmax,.5),                   #fit to means
+     predict(fitAOB2,data.frame(Nitm=seq(xmin,xmax,.5))))
```

where the function ns requires the splines library. The loess fit will vary with the choice of span, as the natural spline fit will vary with the choice of ns.

Covariates can be used as well for more advanced statistical modeling. Indeed, linear and generalized linear (regression) models are similarly well-developed in R. The syntax is slightly unusual, though, since a standard linear regression is launched as follows:

```
> x=seq(-3,3,le=5) # equidispersed regressor
```

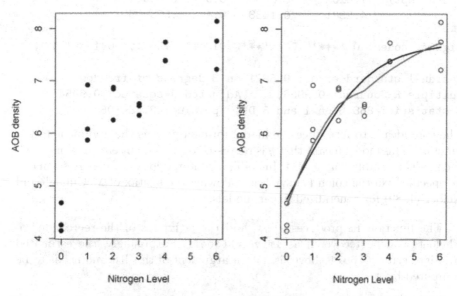

Fig. 1.6. Scatterplot of bacteria abundance (AOB) versus nitrogen levels *(left panel)*. The right panel shows both the natural spline fit *(dark)* with ns=2 and loess fit *(light)* with span=1.25.

```
> y=2+4*x+rnorm(5) # simulated variable
> lm(y~x)

Call:
lm(formula = y ~ x)

Coefficients:
(Intercept)            x
      1.820        4.238

> summary(lm(y~x))

Call:
lm(formula = y ~ x)

Residuals:
       1          2          3          4          5
 0.25219   -0.07421    0.07080   -0.92773    0.67895

Coefficients:
            Estimate Std. Error t value Pr(>|t|)
```

```
(Intercept)    1.8203      0.3050   5.967  0.00942 **
x              4.2381      0.1438  29.472 8.58e-05 ***
---
Signif. codes: 0 '***' .001 '**' .01 '*' 0.05 '.' 0.1 ' ' 1

Residual standard error: 0.6821 on 3 degrees of freedom
Multiple R-Squared: 0.9966,      Adjusted R-squared: 0.9954
F-statistic: 868.6 on 1 and 3 DF,  p-value: 8.58e-05
```

The core idea is to introduce the model formula y~x as the argument to the function. This model means that y is regressed on x. If no intercept is involved, the model is modified as y~x-1. Introducing interactions in the regression can be specified via the colon (:) symbol, following the syntax of McCullagh and Nelder (1989) for generalized linear models.

The function lm produces a list, and the estimates of the regression coefficients can be recovered as lm(y~x)$coeff. Surprisingly, the estimated standard error (0.6821 above) is not an argument of this list and needs to be computed by

```
> out=lm(y~x)
> sqrt(sum(out$res^2)/out$df)
[1] 0.6821
```

rather than via var(out$res), which uses the "wrong" number of degrees of freedom. Note that the package arm (Gelman and Hill, 2006) provides a cleaner output than summary via its display function.

An analysis of variance can be done by recycling output from lm, as in this analysis on the impact of food type on chicken weights:

```
> summary(lm(weight ~ feed, data = chickwts))

Call:
lm(formula = weight ~ feed, data = chickwts)

Residuals:
     Min       1Q   Median       3Q      Max
-123.909  -34.413    1.571   38.170  103.091

Coefficients:
               Estimate Std. Error t value Pr(>|t|)
(Intercept)     323.583     15.834  20.436  < 2e-16 ***
feedhorsebean  -163.383     23.485  -6.957 2.07e-09 ***
feedlinseed    -104.833     22.393  -4.682 1.49e-05 ***
feedmeatmeal    -46.674     22.896  -2.039 0.045567 *
feedsoybean     -77.155     21.578  -3.576 0.000665 ***
```

```
feedsunflower     5.333     22.393    0.238 0.812495
---
Signif. codes: 0 '***' 0.001 '**' 0.01 '*' 0.05 '.' 0.1 ' ' 1

Residual standard error: 54.85 on 65 degrees of freedom
Multiple R-Squared: 0.5417,      Adjusted R-squared: 0.5064
F-statistic: 15.36 on 5 and 65 DF,  p-value: 5.936e-10

> anova(lm(weight ~ feed, data = chickwts))

Analysis of Variance Table

Response: weight
          Df Sum Sq Mean Sq F value   Pr(>F)
feed       5 231129   46226  15.365 5.936e-10 ***
Residuals 65 195556    3009
---
Signif. codes: 0 '***' 0.001 '**' 0.01 '*' 0.05 '.' 0.1 ' ' 1
```

where the first command produces the regression coefficients for each type of food, while the second command evaluates the relevance of the regression model (and concludes positively). When using factor variables, more specific analyzes can be conducted by splitting the degrees of freedom in aov using the option split.

Generalized linear models can be equally well-estimated thanks to the polymorphic function glm. For instance, fitting a binomial generalized linear model to the probability of suffering from diabetes for a woman within the Pima Indian population is done by

```
> glm(formula = type ~ bmi + age, family = "binomial",
+      data = Pima.tr)

Deviance Residuals:
    Min      1Q   Median      3Q      Max
-1.7935  -0.8368  -0.5033  1.0211   2.2531

Coefficients:
             Estimate Std. Error z value Pr(>|z|)
(Intercept) -6.49870    1.17459  -5.533 3.15e-08 ***
bmi          0.10519    0.02956   3.558 0.000373 ***
age          0.07104    0.01538   4.620 3.84e-06 ***
---
Signif. codes: 0 '***' 0.001 '**' 0.01 '*' 0.05 '.' 0.1 ' ' 1

(Dispersion parameter for binomial family taken to be 1)
```

```
    Null deviance: 256.41   on 199   degrees of freedom
Residual deviance: 215.93   on 197   degrees of freedom
AIC: 221.93
```

```
Number of Fisher Scoring iterations: 4
```

concluding with the significance both of the body mass index bmi and the age. Other generalized linear models can be defined by using a different family value whose range is provided by the function family. Note also that link functions different from the intrinsic (and default) link functions (McCullagh and Nelder, 1989) can be specified, as well as a scale factor, as in

```
> glm(y ~ x, family=quasi(var="mu^2", link="log"))
```

where the model corresponds to a quasi-likelihood with link equal to the log function.

Unidimensional and multidimensional time series $(x_t)_t$ can be handled directly by the arima function, following a Box–Jenkins-like analysis,

```
> arima(diff(EuStockMarkets[,1]),order=c(0,0,5))
```

```
Call:
arima(x = diff(EuStockMarkets[, 1]), order = c(0, 0, 5))
```

```
Coefficients:
          ma1      ma2      ma3      ma4      ma5  intercept
       0.0054  -0.0130  -0.0110  -0.0041  -0.0486     2.0692
s.e.   0.0234   0.0233   0.0221   0.0236   0.0235     0.6990
```

```
sigma^2 estimated as 1053:  log likelihood = -9106.23,
aic = 18226.45
```

while more advanced models can be fitted thanks to the function StructTS. Simpler time-series functions can also be used, such as

```
> acf(ldeaths, plot=F) #monthly deaths from bronchitis,
      #emphysema and asthma in the UK, 1974-1979
```

```
Autocorrelations of series 'ldeaths', by lag
```

```
0.0000 0.0833 0.1667 0.2500 0.3333 0.4167 0.5000 0.5833
 1.000  0.755  0.397  0.019 -0.356 -0.609 -0.681 -0.608
0.6667 0.7500 0.8333 0.9167 1.0000 1.0833 1.1667 1.2500
-0.378 -0.013  0.383  0.650  0.723  0.638  0.372  0.009
1.3333 1.4167 1.5000
-0.294 -0.497 -0.586
```

which is less straightforward to analyze than its graphical alternatives

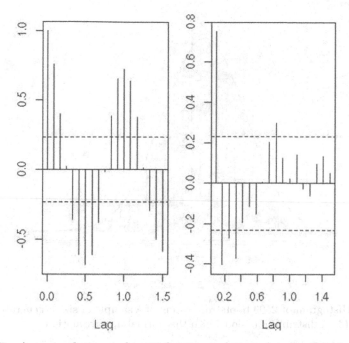

Fig. 1.7. Autocorrelation and partial autocorrelation plots for the ldeaths series, representing monthly deaths from bronchitis, emphysema, and asthma in the UK over the period 1974—1979. The dashed horizontal lines correspond to the significance boundaries for the non-zero terms.

```
> acf(ldeaths)
> acf(ldeaths,type="partial")
```

represented on Figure 1.7 for both the standard and the partial autocorrelations. The standard autocorrelation exhibits quite clearly the seasonal pattern of the deaths.

Lastly, we mention the *bootstrap*. This procedure has many uses (see Efron and Tibshirani, 1993), but here we will mainly illustrate its use as a means of attaching standard errors.

For readers unfamiliar with the notion of the bootstrap, we briefly recall here that this statistical method is based upon the notion that the empirical distribution of a sample $X_1, \ldots, X_n$ converges in n to the true distribution. (The empirical distribution is a discrete distribution that puts a probability $1/n$ on every point X_i of the sample and 0 everywhere else.) The bootstrap procedure then uses the empirical distribution as a substitute for the true distribution to construct variance estimates and confidence intervals. Given

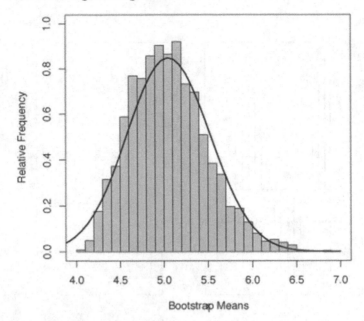

Fig. 1.8. Histogram of 2500 bootstrap means of a sample of size 8 generated from a gamma $\mathcal{G}(4, 1)$ distribution, along with the normal approximation.

that the empirical distribution has a finite but large support made of n^n points, Monte Carlo approximations are almost always necessary to evaluate those quantities, as illustrated in the next example.

For example, if we have a data vector y, we can create a bootstrap sample y^* using the code

```
> ystar=sample(y,replace=T)
```

Figure 1.8 shows a histogram of 2500 bootstrap means `mean(ystar)` based on the sample

$$y = c(4.313, 4.513, 5.489, 4.265, 3.641, 5.106, 8.006, 5.087),$$

along with the normal approximation based on the original sample (i.e., using the empirical mean and empirical variance of y). The sample was in fact drawn from a gamma $\mathcal{G}(4, 1)$ distribution, and we can see on this graph that the bootstrap approximation does capture some of the skewness in the distribution of $\bar{y}$ (indeed, the sample size $n = 8$ is rather small for the Central Limit Theorem to apply). The standard deviation of the sample is 0.4699, while the standard deviation of the 2500 bootstrap means is 0.4368, showing agreement.

One difficulty in implementing the bootstrap technique is that it is not always clear which quantity should be bootstrapped. A treatment of this important topic is outside the scope of this book, and we caution the reader to verify the proper bootstrapping technique in any particular application.

As an example, we apply the bootstrap technique to the simple linear regression example seen previously in this chapter.

Example 1.1. Recall the linear regression

```
> x=seq(-3,3,le=5)              # equidispersed regressor
> y=2+4*x+rnorm(5)              # simulated dependent variable
> lm(y~x)
```

This corresponds to the regression model

$$Y_{ij} = \alpha + \beta x_i + \varepsilon_{ij},$$

where α and β are the unknown intercept and slope, and the ε_{ij} are the iid normal errors. We fit the model with least squares and get $\hat{\alpha} = 1.820$ and $\hat{\beta} = 4.238$.[9] The residuals from the least squares fit are given by

$$\hat{\varepsilon}_{ij} = y_{ij} - \hat{\alpha} - \hat{\beta}x_i,$$

and these are the random variables making the sample that we bootstrap. That is, we create bootstrap samples by resampling the $\hat{\varepsilon}_{ij}$'s, producing a new sample $(\hat{\varepsilon}^*_{ij})_{ij}$ by sampling with replacement from the $\hat{\varepsilon}_{ij}$'s. The bootstrap data are then $y^*_{ij} = y_{ij} + \hat{\varepsilon}^*_{ij}$. This can be implemented with the R code

```
> fit=lm(y~x)                   #fit the linear model
> Rdata=fit$residuals          #get the residuals
> nBoot=2000                    #number of bootstrap samples
> B=array(0,dim=c(nBoot, 2))   #bootstrap array
> for(i in 1:nBoot){            #bootstrap loop
>       ystar=y+sample(Rdata,replace=T)
>       Bfit=lm(ystar~x)
>       B[i,]=Bfit$coefficients
>    }
```

The results of this bootstrap inference are summarized in Figure 1.9, where we provide the histograms of the 2000 bootstrap replicates of both regression coefficients. We can also derive from those replicates confidence intervals on both coefficients (by taking 2.5th and 97.5th percentiles, for example), as well as confidence intervals on predicted values (i.e., for new values of x). For instance, based on the bootstrap sample, the derived 90% confidence intervals are (in our case) $(2.350, 3.416)$ for the intercept α and $(4.099, 4.592)$ for the slope β. ◄

---

[9] When running the same experiment, you will obviously get different numerical values due to the use of a different random seed (see Section 2.1.1).

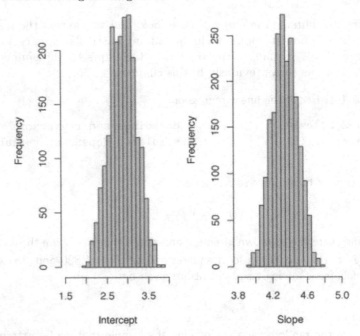

Fig. 1.9. Histogram of 2000 bootstrap intercepts (left) and slopes (right) for the linear regression of Exercise 1.1. The least squares estimate of the intercept is 2.900, and the slope estimate is 4.35.

Exercise 1.7 For the data associated with Figure 1.8:

a. Bootstrap the data and obtain a similar figure based on 1000 bootstrap replications. If the inference is about the 95% of the distribution of $\bar{y}$, $q_{.95}(\bar{y})$, give a bootstrap estimate of this quantity, $\hat{q}_{.95}(\bar{y})$.
b. Construct a bootstrap experiment that provides a 95% confidence interval on $\hat{q}_{.95}(\bar{y})$. (*Hint:* You have to use two levels of bootstrapping to achieve this goal.)

Exercise 1.8 For a dataset simulated as in Example 1.1, compare the bootstrap confidence intervals on both coefficients to the usual ones based on the t-distribution. Comment on the differences.

1.6 Graphical facilities

Another clear advantage of using the R language is that it allows a very rich range of graphical possibilities. Functions such as `plot` and `image` can be customized to a large extent, as described in Venables and Ripley (1999) or

Murrell (2005) (the latter being entirely dedicated to the R graphic abilities).
Even though the default output of plot as for instance in

> plot(faithful)

is not the most enticing, plot is incredibly flexible: To see the number of
parameters involved, you can type par() that delivers the default values of
all those parameters.

⚡ The wealth of graphical possibilities offered by R should be taken advantage of cautiously! That is, good design avoids clutter, small fonts,
  unreadable scale, etc. The recommendations found in Tufte (1990, 2001)
  are thus worth following to avoid horrid outputs like those often found
  in some periodicals! In addition, graphs produced by R usually tend to
  look nicer on the current device than when printed or included in a slide
  presentation. Colors may change, font sizes may turn awkward, separate
  curves may end up overlapping, and so on. In the early stages of working
  with R, and even later, you should thus check that the different outputs
  corresponding to the same graph are satisfactory before closing your R
  session and losing hours of work!!!

Before covering the most standard graphic commands, we start by describing the notion of device that is at the core of those graphic commands. Each
graphical operation sends its outcome to a *device*, which can be a graphical
window (like the one that automatically appears when calling a graphical command for the first time as in the example above) or a file where the graphical
outcome is stored for printing or other uses. Under Unix and Linux OS, launching a new graphical window can be done via X11(), with many possibilities
for customization (such as size, positions, color, etc.). Once a graphical window is created, it is given a device number and can be managed by functions
that start with dev., such as dev.list, dev.set, and others. An important
command is dev.off, which closes the current graphical window. When the
device is a file, it is created by a function that is named after its driver. There
are therefore a postscript, a pdf, a jpeg, and a png function. The complete
list is given by capabilities(). When printing to a file, as in the following
example,

```
> jpeg(file="faith,jpg")
> par(mfrow=c(1,2),mar=c(4,2,2,1))
> hist(faithful[,1],nclass=21,col="grey",main="",
+ xlab=names(faithful)[1])
> hist(faithful[,2],nclass=21,col="wheat",main="",
+ xlab=names(faithful)[2])
> dev.off()
```

closing the sequence with dev.off() is recommended since it completes the
file, which is then saved. If the command jpeg(file="faith,jpg") is repeated, the earlier version of the jpeg file is erased.

Using a line command interface for controlling graphics may seem antiquated, but this is the consequence of the R object-oriented philosophy. In addition, current graphs can be saved to a postscript file using the `dev.copy` and `dev.print` functions. Note that R-produced graphs tend to be large objects, in part because the graphs are not pictures of the current state but instead preserve every action ever taken. For this reason, long series should be thinned down to a few thousand points, images should work with a few hundred pixels, contours should be preferred to images, and jpeg preferred to pdf.

⚡ One of the most frustrating features of R is that the graphical device is not refreshed while a program is executed in the main window. This implies that, if you switch from one terminal to another or if the screen saver starts, the whole or parts of the graph currently on the graphical device will not be visible until the completion of the program. Conversely, refreshing very large graphs will delay the activation of the prompt >.

As already stressed above, `plot` is a highly versatile tool that can be used to represent functional curves and two-dimensional datasets. Colors (chosen by `colors()` or `colours()` out of 650 hues), widths, and types can be calibrated at will and LaTeX-like formulas can be included within the graphs using `expression`; see `plotmath(grDevices)` for a detailed list of the mathematical symbols. Text and legends can be included at a specific point with `locator` (see also `identify`) and `legend`. An example of (relatively simple) output is

```
> plot(as.vector(time(mdeaths)),as.vector(mdeaths),cex=.6,
+ pch=19,xlab="",ylab="Monthly deaths from bronchitis")
> lines(spline(mdeaths),lwd=2,col="chocolate",lty=3)
> ar=arima(mdeaths,order=c(1,0,0))$coef
> lines(as.vector(time(mdeaths))[-1], ar[2]+ar[1]*
+ (mdeaths[-length(mdeaths)]-ar[2]),col="grey",lwd=2,lty=2)
+ title("Splines versus AR(1) predictor")
> ari=arima(mdeaths,order=c(1,0,0),seasonal=list(order=c(1,
+ 0,0),period=12))$coef
> lines(as.vector(time(mdeaths))[-(1:13)],ari[3]+ari[1]*
+ (mdeaths[-c(1:12,72)]-ari[3])+ari[2]*(mdeaths[-(60:72)]-
+ ari[3]),lwd=2,col="steelblue",lty=2)
> title("\n\nand SAR(1,12) predictor")
+ legend(1974,2800,legend=c("spline","AR(1)","SAR(1,12)"),
+ col=c("chocolate","grey","steelblue"),
+ lty=c(3,2,2),lwd=rep(2,3),cex=.5)
```

represented ion Figure 1.10, which compares spline fitting to an AR(1) predictor and to an SAR(1,12) predictor. Note that the seasonal model is doing worse.

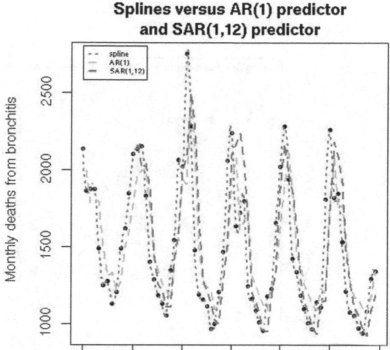

Fig. 1.10. Monthly deaths from bronchitis in the UK over the period 1974—1980 and fits by a spline approximation and two AR predictors.

Another example illustrates the use of the command cumsum, which is particularly handy when checking Monte Carlo convergence, as discussed in the remark box of page 66.

```
> x=rnorm(1)
> for (t in 2:10^3)
+      x=c(x,.09*x[t-1]+rnorm(1))
> plot(x,type="l",xlab="time",ylab="x",lwd=2,lty=2,
+ col="steelblue",ylim=range(cumsum(x)))
> lines(cumsum(x),lwd=2,col="orange3")
```

This four-line program generates a simple AR(1) sequence and plots the original sequence (x_t) along with the cumulated sum sequence,

$$\sum_{i=1}^{t} x_i \, .$$

Note that, due to the high correlation factor (0.9), the cumulated sum is behaving much closer to a random walk.

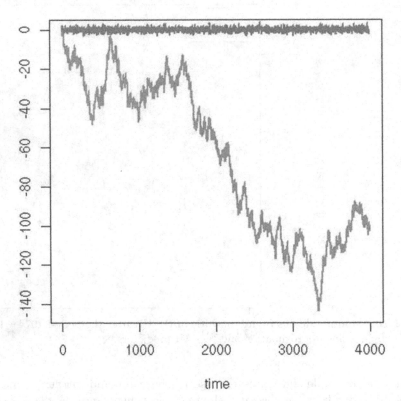

time

Fig. 1.11. Simulated AR(1) sequence *(dotted)* along with its corresponding cumulated sum.

Useful graphical functions include `hist`, for constructing and optionally plotting histograms of datasets; `points`, for adding points on an existing graph; `lines`, for linking points together on an existing graph, as in the above example; `polygon`, for filling the area between two sets of points; `barplot`, for creating barplots; and `boxplot`, for creating boxplots. The two-dimensional representations offered by `image` and `contour` are quite handy when providing likelihood or posterior surfaces, as in Figures 3.5 and 5.2. An instance of using `polygon` is provided by

```
> par(mar=c(2,2,2,2))
> x=matrix(0,ncol=100,nrow=10^4)
> for (t in 2:10^4)
>     x[t,]=x[t-1,]+rnorm(100)*10^(-2)
> plot(seq(0,1,le=10^4),x[,1],ty="n",
+       ylim=range(x),xlab="",ylab="")
> polygon(c(1:10^4,10^4:1)/10^4,c(apply(x,1,max),
+           rev(apply(x,1,min))),col="gold",bor=F)
> polygon(c(1:10^4,10^4:1)/10^4,c(apply(x,1,quantile,.95),
+           rev(apply(x,1,quantile,.05))),col="brown",bor=F)
```

which approximates the range of 100 Brownian motions, as well as a 90% confidence band, represented in Figure 1.12 (see Kendall et al., 2007, and Section 4.5).

⚡ The command `points` is used to add one or several points on a two-dimensional plot. It suffers from a drawback, however, in that the entry is by default a time series. Therefore, calling `points(x)` when x is a two-dimensional vector will plot both points $(1, x_1)$ and $(2, x_2)$ rather than the single point (x_1, x_2). The result will be as expected if x is a two-column matrix, resulting in the points (x_{i1}, x_{i2}) being plotted.

These comments are only here to provide an introduction to the capacities of R. Specific references such as Murrell (2005) need to be consulted to get a complete picture of those capacities!

1.7 Writing new R functions

One of the strengths of R is that new functions and libraries can be created by anyone and then added to Web depositories to continuously enrich the language. These new functions are not distinguishable from the core functions of R, such as `median` or `var`, because those are also written in R. This means their code can be accessed and potentially modified, although it is safer to define new functions. (A few functions are written in C, however, for efficiency.) Learning how to write functions designed for one's own problems is paramount for their resolution, even though the huge collection of available R functions may often contain a function already written for that purpose.

Exercise 1.9 Among the R functions you have met so far, check which ones are written in R by simply typing their name without parentheses, as in mean or var.

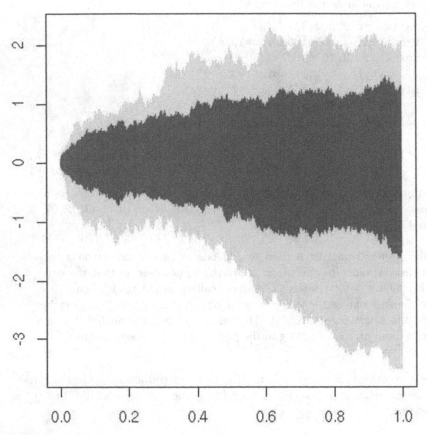

Fig. 1.12. Range of 100 simulated Brownian motions (lighter hue) and 90% confidence band (darker hue).

A function is defined in R by an assignment of the form

```
name=function(arg1[=expr1],arg2[=expr2],...) {
expression
...
expression
value
}
```

where `expression` denotes an R command that uses some of the arguments `arg1`, `arg2`, ... to calculate a value, `value`, that is the outcome of the function. The braces indicate the beginning and the end of the function and the brackets some possible default values for the arguments. Note that producing a value at the end of a function is essential because anything done within a function is local and temporary, and is therefore lost once the function has been exited unless saved in `value` (hence, again, the appeal of `list`). For instance, the following function, named `sqrnt`, implements a version of Newton's method for calculating the square root of `y`:

```
sqrnt=function(y){
    x=y/2
    while (abs(x*x-y) > 1e-10) x=(x+y/x)/2
    x
    }
```

When designing a new R function, it is more convenient to use an external text editor and to store the function under development in an external file, say `myfunction.R`, which can be executed in R as `source("myfunction.R")`. Note also that some external commands can be launched within an R function via the very handy command `system`. This is, for instance, the easiest (if not the most efficient) way to incorporate programs written in other languages (e.g., Fortran, C, Matlab) within R programs.

The fact that R is an interpreted language obviously helps in debugging programs. Indeed, whenever a program is interrupted or aborted, the variables that have been previously defined keep their latest value and can thus be used to assess the cause of the error. This is not always sufficient though, in particular because variables defined within functions are not stored, and a useful tool is the pause command `browser`, which temporarily stops a program to allow the programmer to check the values of all variables at a given point. Further debugging commands are `debug` and `trace`.

⚡ Using an external text editor and an external program file is important for two reasons. First, using the line editor of R is inefficient and close to impossible for complex problems, even though functions can be internally edited by the Unix vi editor as `myfun=vi(myfun)`. Cut-and-paste is just much easier. Second, this forces you to save your functions rather than relying on .RData and .Rhistory, which are not 100% secure.

The expressions used in a function rely on a syntax that is quite similar to those of other programming languages, with conditional statements such as

```
if (expres1) expres2 else expres3
```

where `expres1` is a logical value, and loops such as

```
for (name in expres1) expres2
```

and

```
while (expres4) expres2
```

where `expres1` is a collection of values, as illustrated in Figure 1.13, and `expres4` is a Boolean expression. In particular, Boolean operators can be used within those expressions, including `==` for testing equality, `!=` for testing inequality, `&` for the logical and, `|` for the logical or, and `!` for the logical contradiction.

```
> bool=T;i=0                              separate commands by semicolons
> while(bool==T) {i=i+1; bool=(i<10)}     stop at i = 10
> s=0;x=rnorm(10000)
> system.time(for (i in 1:length(x)){     output sum(x) and
+     s=s+x[i]})[3]                        provide computing time
> system.time(t(rep(1,10000))%*%x)[3]     compare with vector product
> system.time(sum(x))[3]                  compare with sum efficiency
```

Fig. 1.13. Some artificial loops in R.

Exercise 1.10 Explain the difference between the logical operators & and &&, |, ||, and xor.

The operator `if` (and the associated operators `ifelse` and) are some of the rare occurrences where R does not apply to vectors. For vector-valued tests, logical vectors like `(abs(x)>1.96)` can be used as indices of the output vector, like the allocation commands

```
> y[(abs(x)<1.96)]=rep(0,sum(abs(x)>1.96))
> y[(abs(x)>1.96)]=x[(x>1.96)]
```

Since R is an interpreted language, avoiding loops by vectorial programming is generally a good idea, but this may render programs much harder to read. It is therefore extremely useful to include comments within the programs by using the symbol #.

As noted previously, R is fundamentally slower than other languages. Checking both the speed of a program and the reasons for its poor speed can be done using the `system.time` command or the more advanced profiling commands `Rprof` and `Rprofmem` described in the manual. There are, however, ways of speeding up the execution of your programs. First, using faster functions (for example, those already programmed in C; see below) obviously brings improvement. Second, preallocating memory as in `x=double(10^5)` also increases speed. Third (and this is getting way beyond the scope of this introduction!), it is possible to recompile parts of the R package with libraries that are designed for your machine. An example is the `Blas` (basic linear algebra subprogram), which can be optimized using the free library `Atlas` (and lead to improvements by factors from two to five). Details can be found in the

R administration manual. Fourth, and this again requires some programming expertise, you can take advantage of multiple processors, using for instance netWorkSpace (NWS), Rpmi, or snow, developed by Luke Tierney.

While we cannot delve much here into the issue of interfacing R with other languages, we do nonetheless stress that this is an important feature of R that you should investigate, simply because there are problems R is just too slow to handle! Using some routines written in C or Fortran then becomes paramount, without losing the main advantages of R altogether. The easiest way to connect R with external subroutines such as the C executable mycprog.o is to design the corresponding C program to take its input from a file such as mycinput and write its output in another file such as mycouput. In this case, calling

> system("mycprog.o")

within the R program will be sufficient. Obviously, this is a rudimentary type of interfacing and it suffers from two drawbacks, the first one being that repeated access to files is time-consuming as well and the second one being that the C program cannot call R functions this way. A more advanced approach is based on the function .C, which can call C functions with arguments, and the C subroutine call_R, as described for instance in Crawley (2007). The main difficulty with these more advanced techniques is to ensure the compatibility between data types. Section 8.5.2 provides an illustration of a C program being called by an R program in an efficient manner.

1.8 Input and output in R

Large data objects need to be read as values from external files rather than entered during an R session at the keyboard (or by cut-and-paste). Input facilities are simple, but their requirements are fairly strict. In fact, there is a clear presumption that it is possible to modify input files using other tools outside R.

An entire data frame can be read directly with the read.table function. Plain files containing rows of values with a single mode can be downloaded using the scan function, as in

> a=matrix(scan("myfile"),nrow=5,byrow=T)

When data frames have been produced by other statistical software, the library foreign can be used to input those frames in R. For example, the function read.spss allows one to read SPSS data frames.

Conversely, the generic function save can be used to store all R objects in a given file, either in binary or ASCII format. (The alternative function dump is more rudimentary but also useful.) The function write.table is used to export R data frames as ASCII files.

R programs can also be run in BATCH mode, which means that they can run on remote machines. This is particularly relevant when executing highly time-consuming programs that cannot run on your laptop or even personal computer but rather require the power of a multiprocessor mainframe. In this case, a self-contained program, myprogram.R say, can be launched from the operating system (not in R) by

```
R CMD BATCH myprogram.R myprogram.outfile
```

and, provided it does not encounter difficulties or a power outage during its execution, this program will store its outcome in myprogram.outfile.

1.9 Administration of R objects

During an R session, objects are created and stored by name. The command objects() (or, alternatively, ls()) can be used to display, within a directory called the *workspace*, the names of the objects that are currently stored. Individual objects can be deleted with the function rm.

All objects created during an R session (including functions) can be stored permanently in a file for use in future R sessions. At the end of each R session, obtained using the command quit (which can be abbreviated as q), the user is given the opportunity to save all the currently available objects, as in

```
>q()
Save workspace image? [y/n/c]:
```

If the user answers y, the object created during the current session and those saved from earlier sessions are saved in a file called .RData and located in the current directory. When R is called again, it reloads the workspace from this file, which means that the user starts the new session exactly where the old one had stopped. In addition, the entire past command history is stored in the file .Rhistory and can be used in the current session or later by using the command history().

The storage in .RData is specific to the current directory. This means that calling R from another directory will start a new .RData file. This can be useful when running multiple experiments, but if you want to load the workspace from another directory, you should use the function setwd for setting the working directory to this other directory.

1.10 The mcsm package

Since this is primarily a paper book, copying the R code represented on the pages to your computer terminal would be both tedious and time-wasting. We have therefore gathered all the programs and codes of this book within an

R package called mcsm (for Monte Carlo statistical methods) that you should download from CRAN before proceeding to the next chapter. Once downloaded on your computer following the instructions provided on the CRAN Webpage, the package mcsm is loaded into your current R session by library(mcsm). All the functions defined inside the package are then available, and so is a step-by-step reproduction of the examples provided in the book, using the demo command:

```
> demo(Chapter.1)


        demo(Chapter.1)
        ____ ~~~~~~~~~~


Type <Return>   to start :
> # Chapter 1 R commands
>
> # Section 1.3.2
>
> x=matrix(1:4,ncol=3)

> print(x[x>5])
integer(0)

> print(x[1.])
[1] 1

> S=readline(prompt="Type  <Return>   to continue : ")
Type <Return>   to continue :
```

and similarly for the following chapters. Obviously, all commands contained in the demonstrations and all functions defined in the package can be accessed and modified.

↯ Although most steps of the demonstrations are short, some may require longer execution times. If you need to interrupt the demonstration, recall that Ctrl-C is an interruption command.

1.11 Additional exercises

Exercise 1.11 Study the relevance of the attach and assign commands toward handling databases other than .RData.

Exercise 1.12 Construct a vector x that contains integers, real numbers, chains of characters, and several NA missing values. Test for the positions of the missing values using the is.na function. Produce the subvector where all missing values have been eliminated.

Exercise 1.13 Explain the distinctions between the R commands capture.output, dput, dump, save, sink, and write, and illustrate those distinctions in examples of your own.

Exercise 1.14 In conjunction with the discussion of the function var producing NA on page 15, explain the role of the option na.rm in this function and why it does not help with the issue. Describe the role of the option na.action in the functions lm and glm.

Exercise 1.15 Show that, when a is a scalar and x is a vector, match(a,x) is equivalent to min(which(x == a)). Discuss the uses of match and which in the case of the comparison of two vectors. Compare this with the use of %in%.

Exercise 1.16 The Boolean expression x==y does not work well for floating-point numbers in that rounding errors may produce a FALSE answer. Compare it with the functions all.equal and identical.

Exercise 1.17 Compare the execution times of the three equivalent R commands

```
a. y=c();for (t in 1:100) y[t]=exp(t)
b. y=exp(1:100)
c. y=sapply(1:100,exp)
```

using system.time.

Exercise 1.18 Explain why the functions diag, dim, length, and names can be assigned new values (as in diag(m)=pi).

Exercise 1.19 Using the uniform $\mathcal{U}(0,1)$ random generator runif, construct a 2x2 matrix A such that the sum of each row is 1. Show that this property is preserved by a matrix power product, and check whether numerical inaccuracies occur when the power is high enough.

Exercise 1.20 Discuss the relevance of the commands unlist and unclass.

Exercise 1.21 Using the Orange dataset that monitors tree growth versus age for five orange trees, represent the dataset using the command xyplot. Then fit a linear model explaining the circumference by the age via lm. Try using the tree index as an extra covariate.

Exercise 1.22 Here we look at some further applications of the bootstrap.

a. Bootstrap the autocorrelations of Figure 1.7, and compare the bootstrap confidence intervals to the dashed lines given in the plots. (Here the bootstrap sample is taken by resampling the data with replacement.)
b. Bootstrap the spline fit shown in Figure 1.6, and use the results to attach a measure of uncertainty to the spline fit. (Here we bootstrap by resampling the residuals, as in Example 1.1. However, instead of showing histograms of the coefficients, you should present the range of fitted curves on a plot.)

Exercise 1.23 We recall that a Sudoku is a 9x9 grid that is partly filled with numbers between 1 and 9 such that any number between 1 and 9 only appears once in a row, a column, or a 3x3 bloc of the grid. This exercise solves a simple Sudoku grid where there exists a path over the empty entries that fills them one at a time by excluding all possibilities but one. The Sudoku we solve is given by

```
> s=matrix(0,ncol=9,nrow=9)
> s[1,c(6,8)]=c(6,4)
> s[2,c(1:3,8)]=c(2,7,9,5)
> s[3,c(2,4,9)]=c(5,8,2)
> s[4,3:4]=c(2,6)
> s[6,c(3,5,7:9)]=c(1,9,6,7,3)
> s[7,c(1,3:4,7)]=c(8,5,2,4)
> s[8,c(1,8:9)]=c(3,8,5)
> s[9,c(1,7,9)]=c(6,9,1)
```

a. Print the grid on-screen.
b. We define the array `pool=array(TRUE,dim=c(9,9,9))` of possible values for each entry (i,j) of the grid, `pool[i,j,k]` being FALSE if the value k can be excluded. Give the R code that updates `pool` for the entries already filled.
c. If i is an integer between 1 and 81, explain the meaning of `s[i]`.
d. Show that, for a given entry (a,b), the indices of the integers in the same 3x3 box as (a,b) are defined by

```
boxa=3*trunc((a-1)/3)+1
boxa=boxa:(boxa+2)
boxb=3*trunc((b-1)/3)+1
boxb=boxb:(boxb+2)
```

e. Deduce that values at an entry (a,b) that is not yet determined can be excluded by

```
for (u in (1:9)[pool[a,b,]])
  pool[a,b,u]=(sum(u==s[a,])+sum(u==s[,b])+
             sum(u==s[boxa,boxb]))==0
```

and that certain entries correspond to

```
if (sum(pool[a,b,])==1)  s[i]=(1:9)[pool[a,b,]]
```

f. Solve the grid above by a random exploration of entries (a,b) that continues as long as `sum(s==0)>0`.

⚡ If you ever attempt to apply this program on an arbitrary Sudoku grid, be aware that it may run forever since the harder Sudokus do not allow this logical and deterministic filling but instead require several scenarios in parallel be followed.

2

Random Variable Generation

"It has long been an axiom of mine that the little things are infinitely the most important."

Arthur Conan Doyle
A Case of Identity

Reader's guide

In this chapter, we present practical techniques that can produce random variables from both standard and nonstandard distributions by using a computer program. Given the availability of a uniform generator in R, as explained in Section 2.1.1, we do not deal with the specific production of uniform random variables. The most basic techniques relate the distribution to be simulated to a uniform variate by a transform or a particular probabilistic property, as in Section 2.2, while the most generic one is a simulation version of the trial-and-error method, described in Section 2.3 under the name of the Accept–Reject method. In all cases, the methods rely on the availability of sequences of independent uniform generations that are provided by the resident R generator, runif.

C.P. Robert, G. Casella, *Introducing Monte Carlo Methods with R*, Use R,
DOI 10.1007/978-1-4419-1576-4_2, © Springer Science+Business Media, LLC 2010

2.1 Introduction

The methods developed in this book and summarized under the denomination of *Monte Carlo methods* mostly rely on the possibility of producing (with a computer) a supposedly endless flow of random variables for well-known or new distributions. Such a simulation is, in turn, based on the production of uniform random variables on the interval $(0, 1)$. Although we are not directly concerned with the *mechanics* of producing such uniform random variables, because existing uniform generators can be considered as "perfect", we will completely rely on those generators to produce other random variables. In a sense, the uniform distribution $\mathcal{U}_{[0,1]}$ provides the basic probabilistic representation of randomness on a computer and the generators for all other distributions do require a sequence of uniform variables to be simulated.

As already pointed out in Section 1.4 of Chapter 1, R has a large number of built-in functions that will generate the standard random variables listed in Table 1.1. For instance,

```
> rgamma(3,2.5,4.5)
```

produces three independent generations from a $\mathcal{G}(5/2, 9/2)$ distribution with all due guarantees of representing this distribution. It is therefore counterproductive, inefficient, and even dangerous to generate from those standard distributions using anything but the resident R generators. The principles developed in the following sections are, however, essential to deal with less standard distributions that are not built into R.

2.1.1 Uniform simulation

The basic uniform generator in R is the function `runif`, whose only required entry is the number of values to be generated. The other optional parameters are `min` and `max`, which characterize the bounds of the interval supporting the uniform. (The default is `min=0` and `max=1`.) For instance,

```
> runif(100, min=2, max=5)
```

will produce 100 random variables distributed uniformly between 2 and 5.

Strictly speaking, all the methods we will see (and this includes `runif`) produce *pseudo-random numbers* in that there is no randomness involved—based on an initial value u_0 of a uniform $\mathcal{U}(0, 1)$ sequence and a transformation D, the uniform generator produces a sequence $(u_i) = (D^i(u_0))$ of values in $(0, 1)$—but the outcome has the same *statistical properties* as an iid sequence. Further details on the random generator of R are provided in the on-line help on RNG.

While extensive testing of this function has been undertaken to make sure it does produce uniform variates for all purposes (see, e.g., Robert and Casella,

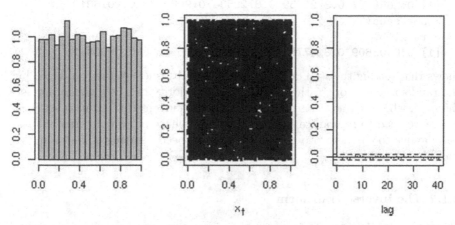

Fig. 2.1. Histogram *(left)*, pairwise plot *(center)*, and estimated autocorrelation function *(right)* of a sequence of 10^4 uniform random numbers generated by runif.

2004, Chapter 2), a quick check on the properties of this uniform generator is to look at an histogram of the X_i's, a plot of the pairs (X_i, X_{i+1}), and the estimated autocorrelation function, as any random variable generator does suffer from a residual autocorrelation and good algorithms will reduce this to a negligible value. The R code used to produce the output in Figure 2.1 is

```
> Nsim=10^4              #number of random numbers
> x=runif(Nsim)
> x1=x[-Nsim]            #vectors to plot
> x2=x[-1]               #adjacent pairs
> par(mfrow=c(1,3))
> hist(x)
> plot(x1,x2)
> acf(x)
```

and shows that **runif** is apparently acceptable for this casual evaluation.

As pointed out in the previous remark, **runif** does not involve randomness per se. Producing **runif(Nsim)** is better described as a deterministic sequence based on a random starting point. An extreme illustration of this fact is obtained through the R function **set.seed**, which uses its single integer argument to set as many seeds as required. For instance,

```
> set.seed(1)
> runif(5)
[1] 0.2655087 0.3721239 0.5728534 0.9082078 0.2016819
> set.seed(1)
> runif(5)
```

```
[1] 0.2655087 0.3721239 0.5728534 0.9082078 0.2016819
> set.seed(2)
> runif(5)
[1] 0.0693609 0.8177752 0.9426217 0.2693818 0.1693481
```

shows that setting the seed determines all the subsequent values produced by
the random generator. In the overwhelming majority of cases, we do not set
the seed, which is then chosen according to the current time. But in settings
where we need to reproduce the exact same sequence of random simulations,
for example to compare two procedures or two speeds, setting a fixed value of
the seed is reasonable.

2.1.2 The inverse transform

There is a simple, sometimes useful transformation, known as the *probability
integral transform*, that allows us to transform any random variable into a
uniform random variable and, more importantly, vice versa. For example, if
X has density f and cdf F, then we have the relation

$$F(x) = \int_{-\infty}^{x} f(t)\, dt,$$

and if we set $U = F(X)$, then U is a random variable distributed from a
uniform $\mathcal{U}(0,1)$. This is because

$$P(U \leq u) = P[F(X) \leq F(x)] = P[F^{-1}(F(X)) \leq F^{-1}(F(x))] = P(X \leq x),$$

where we have assumed that F has an inverse. This assumption can be relaxed
(see Robert and Casella, 2004, Section 2.1) but holds for most continuous
distributions.

Exercise 2.1 For an arbitrary random variable X with cdf F, define the generalized inverse of F by.

$$F^-(u) = \inf \{x;\, F(x) \geq u\}.$$

Show that if $U \sim \mathcal{U}(0,1)$, then $F^-(U)$ is distributed like X.

Example 2.1. If $X \sim \mathcal{E}xp(1)$, then $F(x) = 1 - e^{-x}$. Solving for x in $u = 1 - e^{-x}$
gives $x = -\log(1-u)$. Therefore, if $U \sim \mathcal{U}_{[0,1]}$, then

$$X = -\log U \sim \mathcal{E}xp(1)$$

(as U and $1 - U$ are both uniform). The corresponding R code

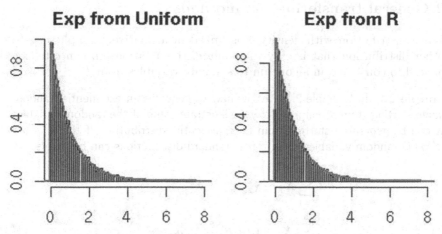

Fig. 2.2. Histograms of exponential random variables using the inverse transform *(right)* and using the R command `rexp` *(left)*, with the $\mathcal{E}xp(1)$ density on top.

```
> Nsim=10^4                  #number of random variables
> U=runif(Nsim)
> X=-log(U)                  #transforms of uniforms
> Y=rexp(Nsim)               #exponentials from R
> par(mfrow=c(1,2))          #plots
> hist(X,freq=F,main="Exp from Uniform")
> hist(Y,freq=F,main="Exp from R")
```

compares the output from the probability inverse transform with the output from rexp. The fits of both histograms to their exponential limit are not distinguishable in Figure 2.2. ◀

The generation of uniform random variables is therefore a key determinant of the behavior of simulation methods for other probability distributions since those distributions can be represented as a deterministic transformation of uniform random variables.

Exercise 2.2 Two distributions that have explicit forms of the cdf are the logistic and Cauchy distributions. Thus, they are well-suited to the inverse transform method. For each of the following, verify the form of the cdf and then generate 10,000 random variables using the inverse transform. Compare your program with the built-in R functions `rlogis` and `rcauchy`, respectively:

a. Logistic pdf: $f(x) = \frac{1}{\beta} \frac{e^{-(x-\mu)/\beta}}{[1+e^{-(x-\mu)/\beta}]^2}$, cdf: $F(x) = \frac{1}{1+e^{-(x-\mu)/\beta}}$.

b. Cauchy pdf: $f(x) = \frac{1}{\pi\sigma} \frac{1}{1+\left(\frac{x-\mu}{\sigma}\right)^2}$, cdf: $F(x) = \frac{1}{2} + \frac{1}{\pi}\arctan((x-\mu)/\sigma)$.

2.2 General transformation methods

When a distribution with density f is linked in a relatively simple way to another distribution that is easy to simulate, this relationship can often be exploited to construct an algorithm to simulate variables from f.

Example 2.2. In Example 2.1, we saw how to generate an exponential random variable starting from a uniform. Now we illustrate some of the random variables that can be generated starting from an exponential distribution. If the X_i's are iid $\mathcal{E}xp(1)$ random variables, then three standard distributions can be derived as

$$Y = 2\sum_{j=1}^{\nu} X_j \sim \chi^2_{2\nu}, \qquad \nu \in \mathbb{N}^*,$$

(2.1)
$$Y = \beta \sum_{j=1}^{a} X_j \sim \mathcal{G}(a, \beta), \qquad a \in \mathbb{N}^*,$$

$$Y = \frac{\sum_{j=1}^{a} X_j}{\sum_{j=1}^{a+b} X_j} \sim \mathcal{B}e(a, b), \qquad a, b \in \mathbb{N}^*,$$

where $\mathbb{N}^* = \{1, 2, \ldots\}$. For example, to generate χ^2_6 random variables, we could use the R code

```
> U=runif(3*10^4)
> U=matrix(data=U,nrow=3)  #matrix for sums
> X=-log(U)                #uniform to exponential
> X=2* apply(X,2,sum)      #sum up to get chi squares
```

Obviously, this is not nearly as efficient as calling rchisq, as can be checked by the R code

```
> system.time(test1());system.time(test2())
   user   system elapsed
  0.104   0.000   0.107
   user   system elapsed
  0.004   0.000   0.004
```

where test1 corresponds to the R code above and test2 to its substitution by X=rchisq(10^4,df=6). ◀

   Many other derivations of standard distributions are possible when taking advantage of existing probabilistic properties, as shown in Exercise 2.12.

   ⚡ These transformations are quite simple to use and hence will often be a favorite in our illustrations. However, there are limits to their usefulness, both in the scope of variables that can be generated that way (think, for instance, of a chi-squared distribution with a noneven number of degrees of freedom) and efficiency of generation. For any specific distribution, efficient algorithms have been developed. Thus, if R has a distribution built in, it is almost always worth using, as shown by Example 2.2. Moreover, the transformation method described above cannot reach all distributions; for example, we cannot get a standard normal.

2.2.1 A normal generator

One way to achieve normal random variable simulation using a transform is with the Box–Muller algorithm, devised for the generation of $\mathcal{N}(0,1)$ variables.

Example 2.3. If U_1 and U_2 are iid $\mathcal{U}_{[0,1]}$, the variables X_1 and X_2 defined by

$$X_1 = \sqrt{-2\log(U_1)}\,\cos(2\pi U_2)\,, \qquad X_2 = \sqrt{-2\log(U_1)}\,\sin(2\pi U_2)\,,$$

are then iid $\mathcal{N}(0,1)$ by virtue of a simple change of variable argument. Note that this is *not* the generator implemented in R, which uses by default the probability inverse transform, based on a very accurate representation of the normal cdf inverse qnorm (up to 16 digits!). (It is, however, possible, if not recommended, to switch the normal generator to the Box–Muller (or even to the Kinderman-Ramage) version via the RNG function.) ◄

In comparison with (crudely) approximative algorithms based on the Central Limit Theorem (CLT), the Box–Muller algorithm is exact, producing two normal random variables from two uniform random variables, the only drawback (in speed) being the necessity of calculating transcendental functions such as log, cos, and sin.

Exercise 2.3 An antiquated generator for the normal distribution is:

> Generate $U_1,\ldots,U_{12} \sim \mathcal{U}[-1/2, 1/2]$
> Set $Z = \sum_{i=1}^{12} U_i$

the argument being that the CLT normality is sufficiently accurate with 12 terms.

a. Show that $\mathbb{E}[Z] = 0$ and $\mathrm{var}(Z) = 1$.
b. Using histograms, compare this CLT-normal generator with the Box–Muller algorithm. Pay particular attention to tail probabilities.
c. Compare both of the generators in part a. with rnorm.

Note that this exercise does not suggest *using* the CLT for normal generations! This is a very poor approximation indeed.

The simulation of a multivariate normal distribution $\mathcal{N}_p(\mu, \Sigma)$, where Σ is a $p \times p$ symmetric and positive-definite matrix, can be derived from the generic rnorm generator in that using a Cholesky decomposition of Σ (that is, $\Sigma = AA^{\mathrm{T}}$) and taking the transform by A of an iid normal vector of dimension p leads to a $\mathcal{N}_p(0, \Sigma)$ normal vector. There is, however, an R package that replicates those steps, called `rmnorm` and available from the `mnormt` library (Genz and Azzalini, 2009). This library also allows computation of the probability of hypercubes via the function `sadmvn`, as in

```
> sadmvn(low=c(1,2,3),upp=c(10,11,12),mean=rep(0,3),var=B)
[1] 9.012408e-05
attr(,"error")
[1] 1.729111e-08
```

where B is a positive-definite matrix. This is quite useful since the analytic derivation of this probability is almost always impossible.

Exercise 2.4 Given a 3×3 matrix `Sigma`:

a. Show that `Sigma=cov(matrix(rnorm(30),nrow=10))` defines a proper covariance matrix.
b. Show that setting `A=t(chol(Sigma))` leads to a simulation from $\mathcal{N}_p(0, \Sigma)$ by using the command `x=A%*%rnorm(3)`.
c. Compare the execution times of this approach and `rmnorm` when simulating one vector and 100 vectors.

2.2.2 Discrete distributions

We next turn to the generation of discrete random variables, where we have an "all-purpose" algorithm. Again using the inverse transform principle of Section 2.1.2, we can indeed construct a generic algorithm that will formally work for any discrete distribution.

To generate $X \sim P_\theta$, where P_θ is supported by the integers, we can calculate—once for all, assuming we can store them—the probabilities

$$p_0 = P_\theta(X \leq 0), \quad p_1 = P_\theta(X \leq 1), \quad p_2 = P_\theta(X \leq 2), \quad \ldots,$$

and then generate $U \sim \mathcal{U}_{[0,1]}$ and take

$$X = k \text{ if } p_{k-1} < U < p_k.$$

Example 2.4. To generate $X \sim \mathcal{B}in(10, .3)$, the probability values are obtained by `pbinom(k,10,.3)` as

$$p_0 = 0.028, \quad p_1 = 0.149, \quad p_2 = 0.382, \ldots, p_{10} = 1,$$

and to generate $X \sim \mathcal{P}(7)$, take

$$p_0 = 0.0009, \quad p_1 = 0.0073, \quad p_2 = 0.0296, \ldots,$$

the sequence being stopped when it reaches 1 with a given number of decimals. (For instance, $p_{20} = 0.999985$.) ◄

Specific algorithms are usually more efficient (as shown in Example 2.5), but it is mostly because of the storage problem. We can often improve on the algorithm above by a judicious choice of what probabilities we compute first. For example, if we want to generate random variables from a Poisson distribution with mean $\lambda = 100$, the algorithm above is woefully inefficient. This is because we expect most of our observations to be in the interval $\lambda \pm 3\sqrt{\lambda}$ (recall that λ is both the mean and the variance for the Poisson distribution), and for $\lambda = 100$ this interval is $(70, 130)$. Thus, starting at 0 will almost always produce 70 tests of whether or not $p_{k-1} < U < p_k$ that are useless because they will almost certainly be rejected. A first remedy is to "ignore" what is outside of a highly likely interval such as $(70, 130)$ in the current example, as

$$P(X < 70) + P(X > 130) = 0.00268.$$

Formally, we should find a lower and an upper bound to make this probability small enough, but informally $\pm 3\sigma$ works fine.

Example 2.5. Here is an R code that can be used to generate Poisson random variables for large values of lambda. The sequence t contains the integer values in the range around the mean.

```
> Nsim=10^4; lambda=100
> spread=3*sqrt(lambda)
> t=round(seq(max(0,lambda-spread),lambda+spread,1))
> prob=ppois(t, lambda)
> X=rep(0,Nsim)
> for (i in 1:Nsim){
+    u=runif(1)
+    X[i]=t[1]+sum(prob<u) }
```

The last line of the program checks to see what interval the uniform random variable fell in and assigns the correct Poisson value to X. See Exercise 2.14 for other distributions. ◄

A more formal remedy to the inefficiency of starting the cumulative probabilities at p_0 is to start instead from the mode of the discrete distribution P_θ and to explore the neighboring values until the cumulative probability is 1 up to an approximation error. The p_k's are then indexed by the visited values rather than by the integers, but the validity of the method remains complete.

Specific algorithms exist for almost any distribution and are often quite fast. Thus, we once again stress that, if R has the distribution that you are interested in, the wisest course is to use it. Once again, the comparison of the code of Example 2.5 with the resident rpois shows how inefficient this simple implementation can be. If test3 corresponds to the above and test4 to rpois, the execution times are given by

```
> system.time(test3()); system.time(test4())
   user  system elapsed
  0.436   0.000   0.435
   user  system elapsed
  0.008   0.000   0.006
```

However, R does not handle every distribution that we will need, so approaches such as the above can be useful. See Exercise 2.15 for some specific algorithms.

2.2.3 Mixture representations

It is sometimes the case that a probability distribution can be naturally represented as a *mixture distribution*; that is, we can write it in the form

$$(2.2) \qquad f(x) = \int_{\mathcal{Y}} g(x|y)p(y) \, dy \qquad \text{or} \qquad f(x) = \sum_{i \in \mathcal{Y}} p_i \, f_i(x) \,,$$

depending on whether the auxiliary space $\mathcal{Y}$ is continuous or discrete, where g and p are standard distributions that can be easily simulated. To generate a random variable X using such a representation, we can first generate a variable Y from the mixing distribution and then generate X from the selected conditional distribution. That is,

if $y \sim p(y)$ and $X \sim f(x|y)$, then $X \sim f(x)$ (if continuous);
if $\gamma \sim P(\gamma = i) = p_i$ and $X \sim f_\gamma(x)$, then $X \sim f(x)$ (if discrete).

For instance, we can write Student's t density with ν degrees of freedom $\mathcal{T}_\nu$ as a mixture, where

$$X|y \sim \mathcal{N}(0, \nu/y) \quad \text{and} \quad Y \sim \chi_\nu^2.$$

Generating from a $\mathcal{T}_\nu$ distribution could then amount to generating from a χ_ν^2 distribution and then from the corresponding normal distribution. (Obviously, using rt is slightly more efficient, as you can check via system.time.)

Example 2.6. If X is a negative binomial random variable, $X \sim \mathcal{N}eg(n, p)$, then X has the mixture representation

$$X|y \sim \mathcal{P}(y) \quad \text{and} \quad Y \sim \mathcal{G}(n, \beta),$$

where $\beta = (1 - p)/p$. The following R code generates from this mixture

```
> Nsim=10^4
> n=6;p=.3
> y=rgamma(Nsim,n,rate=p/(1-p))
> x=rpois(Nsim,y)
> hist(x,main="",freq=F,col="grey",breaks=40)
> lines(1:50,dnbinom(1:50,n,p),lwd=2,col="sienna")
```

and produces Figure 2.3, where the fit to the negative binomial pdf is shown as well.   ◀

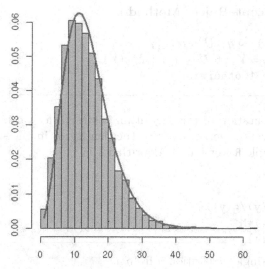

Fig. 2.3. Histogram of 10^4 negative binomial $\mathcal{N}eg(6, .3)$ random variables generated from the mixture representation along with the probability function.

2.3 Accept–reject methods

There are many distributions for which the inverse transform method and even general transformations will fail to be able to generate the required random variables. For these cases, we must turn to *indirect* methods; that is, methods in which we generate a candidate random variable and only accept it subject to passing a test. As we will see, this class of methods is extremely powerful and will allow us to simulate from virtually any distribution.

These so-called *Accept–Reject methods* only require us to know the functional form of the density f of interest (called the *target density*) up to a multiplicative constant. We use a simpler (to simulate) density g, called the *instrumental or candidate density*, to generate the random variable for which the simulation is actually done. The only constraints we impose on this candidate density g are that

(i). f and g have compatible supports (i.e., $g(x) > 0$ when $f(x) > 0$).
(ii). There is a constant M with $f(x)/g(x) \leq M$ for all x.

In this case, X can be simulated as follows. First, we generate $Y \sim g$ and, independently, we generate $U \sim \mathcal{U}_{[0,1]}$. If

$$U \leq \frac{1}{M} \frac{f(Y)}{g(Y)},$$

then we set $X = Y$. If the inequality is not satisfied, we then discard Y and U and start again. Succinctly, the algorithmic representation of the Accept–Reject method is as follows:

Algorithm 1 Accept–Reject Method

1. Generate $Y \sim g$, $U \sim \mathcal{U}_{[0,1]}$;
2. Accept $X = Y$ if $U \leq f(Y)/Mg(Y)$;
3. Return to 1 otherwise.

The R implementation of this algorithm is straightforward: If `randg` is a function that delivers generations from the density g, in the same spirit as `rnorm` or `rt`, a simple R version of Algorithm 1 is

```
> u=runif(1)*M
> y=randg(1)
> while (u>f(y)/g(y)){
+    u=runif(1)*M
+    y=randg(1)}
```

which produces a single generation y from f.

Why does this method work? A straightforward probability calculation shows that the cdf of the accepted random variable, $P(Y \leq x | U \leq f(Y)/\{Mg(Y)\})$, is exactly the cdf of X. That is,

$$
\begin{aligned}
P(Y \leq x | U \leq f(Y)/\{Mg(Y)\}) &= \frac{P(Y \leq x, U \leq f(Y)/\{Mg(Y)\})}{P(U \leq f(Y)/\{Mg(Y)\})} \\
&= \frac{\int_{-\infty}^{x} \int_0^{f(y)/\{Mg(y)\}} \mathrm{d}u \, g(y) \, \mathrm{d}y}{\int_{-\infty}^{\infty} \int_0^{f(y)/\{Mg(y)\}} \mathrm{d}u \, g(y) \, \mathrm{d}y} \\
&= \frac{\int_{-\infty}^{x} [f(y)/\{Mg(y)\}] \, g(y) \, \mathrm{d}y}{\int_{-\infty}^{\infty} [f(y)/\{Mg(y)\}] \, g(y) \, \mathrm{d}y} \\
&= \frac{\int_{-\infty}^{x} f(y) \, \mathrm{d}y}{\int_{-\infty}^{\infty} f(y) \, \mathrm{d}y} = P(X \leq x),
\end{aligned}
$$

where we use the fact that the uniform integral is equal to its upper limit. Despite simulating only from g, the output of this algorithm is thus exactly distributed from f.

⚡ The Accept–Reject method is applicable in any dimension, provided g is a density over the same space as f.

Note the cancellation of the $g(y)$'s and the M's in the integrals above. It also follows from this representation that we do not need to be concerned about normalizing constants. As long as we know f/g up to a constant, $f/g \propto \tilde{f}/\tilde{g}$, the algorithm can be implemented if an upper bound $\widetilde{M}$ can be found on $\tilde{f}/\tilde{g}$. (The missing constants actually get absorbed into M.)

Exercise 2.5 Show that the probability of acceptance in an Accept–Reject algorithm with upper bound M on the density ratio f/g is $1/M$. Show that the expected value of the acceptance rate, $\mathbb{E}[I(U < \tilde{f}/\tilde{M}\tilde{g})]$, can be used to compute the missing constant in f/g.

As stressed by this exercise, the probability of acceptance is $1/M$ *only if* the normalizing constants are known. Otherwise, since the missing constants do get absorbed into $\tilde{M}$, $1/\tilde{M}$ is not the probability of acceptance.

Example 2.7. Example 2.2 did not provide a general algorithm to simulate beta $\mathcal{Be}(\alpha, \beta)$ random variables. We can, however, construct a toy algorithm based on the Accept–Reject method, using as the instrumental distribution the uniform $\mathcal{U}_{[0,1]}$ distribution when both α and β are larger than 1. (The generic rbeta function does not impose this restriction.)

The upper bound M is then the maximum of the beta density, obtained for instance by optimize (or its alias optimise):

```
> optimize(f=function(x){dbeta(x,2.7,6.3)},
+ interval=c(0,1),max=T)$objective
[1] 2.669744
```

Since the candidate density g is equal to one, the proposed value Y is accepted if $M \times U < f(Y)$, that is, if $M \times U$ is under the beta density f at that realization. Note that generating $U \sim \mathcal{U}_{[0,1]}$ and multiplying by M is equivalent to generating $U \sim \mathcal{U}_{[0,M]}$. For $\alpha = 2.7$ and $\beta = 6.3$, an alternative R implementation of the Accept–Reject algorithm is

```
> Nsim=2500
> a=2.7;b=6.3
> M=2.67
> u=runif(Nsim,max=M)     #uniform over (0,M)
> y=runif(Nsim)           #generation from g
> x=y[u<dbeta(y,a,b)]     #accepted subsample
```

and the left panel in Figure 2.4 shows the results of generating 2500 pairs (Y, U) from $\mathcal{U}_{[0,1]} \times \mathcal{U}_{[0,M]}$. The black dots $(Y, Ug(Y))$ that fall under the density f are those for which we accept $X = Y$, and we reject the grey dots $(Y, Ug(Y))$ that fall outside. It is again clear from this graphical representation that the black dots are uniformly distributed over the area under the density f. Since the probability of acceptance of a given simulation is $1/M$ (Exercise 2.5), with $M = 2.67$ we accept approximately $1/2.67 = 37\%$ of the values. ◄

In the implementation of the Accept–Reject algorithm above, the total number of attempts Nsim is fixed, which means that the number of accepted values is a binomial random variable with probability $1/M$. Instead, in most cases, the number of accepted values is fixed, but this implementation can nonetheless be exploited as in

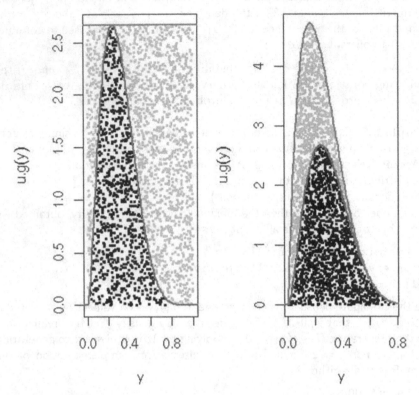

Fig. 2.4. Generation of beta random variables $X \sim \mathcal{Be}(2.7, 6.3)$: Using the Accept–Reject algorithm, 2500 (Y, U) proposals were generated from g and $\mathcal{U}_{[0,M]}$, respectively, and the points $(Y, Ug(Y))$ were represented with grey dots. In the left panel, $Y \sim \mathcal{U}_{[0,1]}$, and 36% of the candidate random variables were accepted and represented with black dots. In the right panel, $Y \sim \mathcal{Be}(2, 6)$ and 58% of the simulated values were accepted (and similarly represented with black dots). In both panels, f and Mg are also plotted.

```
> x=NULL
> while (length(x)<Nsim){
+     y=runif(Nsim*M)
+     x=c(x,y[runif(Nsim*M)*M<dbeta(y,a,b)])}
> x=x[1:Nsim]
```

(Note that using y=u=runif(Nsim*M) in the program would produce a bias, as y and u would then take the same values.) Simulating Nsim*M proposals from the start reduces the number of calls to while since this is the expected number of proposals (Exercise 2.5).

Exercise 2.6 Compare the execution times of the two proposed implementations of the Accept–Reject algorithm, as well as alternatives simulating `Nsim*Nprop` proposals at once when `Nprop` varies.

↯ Some key properties of the Accept–Reject algorithm, which should always be considered when using it, are the following:

1. Only the ratio f/M is needed, so the algorithm does not depend on the normalizing constant.
2. The bound $f \le Mg$ need not be tight; the algorithm remains valid (if less efficient) when M is replaced with any larger constant.
3. The probability of acceptance is $1/M$, so M should be as small as possible for a given computational effort.

The efficiency of a given Accept–Reject algorithm can be measured in terms of its acceptance probability since the higher this probability, the fewer wasted simulations from g. (In absolute terms, this must be weighted down by the computational cost of producing a value from g since otherwise the best choice for g would be f!)

Example 2.8. (Continuation of Example 2.7) Consider instead simulating $Y \sim \mathcal{Be}(2,6)$ as a proposal distribution. This choice of g is acceptable since

```
> optimize(f=function(x){dbeta(x,2.7,6.3)/dbeta(x,2,6)},
+ max=T,interval=c(0,1))$objective
1.671808
```

This modification of the proposal thus leads to a smaller value of M and a correspondingly higher acceptance rate of 58% than with the uniform proposal. The right panel of Figure 2.4 shows the outcome of the corresponding Accept–Reject algorithm and illustrates the gain in efficiency brought by simulating points in a smaller set.   ◄

Exercise 2.7 Show formally that, for the ratio f/g to be bounded when f is a $\mathcal{Be}(\alpha, \beta)$ density and g is a $\mathcal{Be}(a, b)$ density, we must have both $a \le \alpha$ and $b \le \beta$. Deduce that the best choice for a and b among the integer values is $a = \lfloor \alpha \rfloor$ and $b = \lfloor \beta \rfloor$.

As shown by Example 2.8, some optimization of the Accept–Reject algorithm is possible by choosing the candidate density g in a parametric family and by then determining the value of the parameter that minimizes the bound M.

Exercise 2.8 Consider using the Accept–Reject algorithm to generate a $\mathcal{N}(0,1)$ random variable from a double-exponential distribution $\mathcal{L}(\alpha)$, with density $g(x|\alpha) = (\alpha/2)\exp(-\alpha|x|)$ as a candidate.

a. Show that

$$\frac{f(x)}{g(x|\alpha)} \leq \sqrt{\frac{2}{\pi}}\,\alpha^{-1}e^{\alpha^2/2}$$

   and that the minimum of this bound (in α) is attained for $\alpha = 1$.

b. Show that the probability of acceptance is then $\sqrt{\pi/2e} = .76$ and deduce that, to produce one normal random variable, this Accept–Reject algorithm requires on average $1/.76 \approx 1.3$ uniform variables.

c. Show that $\mathcal{L}(\alpha)$ can be generated by the probability inverse transform, and compare this algorithm with the Box–Muller algorithm of Example 2.3 in terms of execution time.

It may sometimes happen that the complexity of the optimization is very expensive in terms of analysis or computing time. In the first case, the construction of the optimal algorithm should still be undertaken when the algorithm is to be subjected to intensive use. This is, for instance, the case for most random generators in R, as can be checked by `help`. In the second case, it is most often preferable to explore the use of another family of instrumental distributions g. (See Exercise 2.22.)

One particular application of the Accept–Reject algorithm has found a niche in population genetics and is called ABC, following the denomination proposed by Pritchard et al. (1999). The core version of this algorithm is an Accept–Reject algorithm fitted for Bayesian problems, where a posterior distribution $\pi(\theta|x_0) \propto \pi(\theta)f(x_0|\theta)$ is to be simulated for a likelihood function $f(x|\theta)$ that is not available but can be simulated. The ABC algorithm then generates values from the prior and from the likelihood until the simulated observation is equal to the original observation x_0:

Repeat

   Generate $\theta \sim \pi(\theta)$ and $X \sim f(x|\theta)$

until $X = x_0$

Exercise 2.9 Prove that the conditional probability of acceptance in the loop above is $f(x_0|\theta)$, and deduce that the distribution of the accepted θ is $\pi(\theta|x_0)$.

This algorithm is thus valid, but it only applies in settings where $\pi(\theta)$ is a proper prior and where $P_\theta(X = x_0)$ has a positive probability of occurring. Even in population genetics where X is a discrete random variable, the size of the state-space is often such that this algorithm cannot be implemented. The proposal of Pritchard et al. (1999) is then to replace the *exact* acceptance condition $X = x_0$ with an *approximate* condition $d(X, x_0) < \epsilon$, where d is a distance and ϵ a tolerance level. While unavoidable, this approximation step makes the ABC method difficult to recommend on a general basis, even though more recent works rephrase it in a non-parametric framework that aims at approximating the likelihood function $f(x|\theta)$ (Beaumont et al., 2002).

One criticism of the Accept–Reject algorithm is that it generates "useless" simulations from the proposal g when rejecting, even those necessary to validate the output as being generated from the target f. We will see in Chapter 3 how the method of importance sampling (Section 3.3) can be used to bypass this problem.

2.4 Additional exercises

Exercise 2.10 The vector `randu` is a historical reminder of how wrong a random generator can get. It consists of 400 rows of three consecutive values produced by a former VAX random generator called RANDU.

  a. Produce a random sample by taking all columns of `randu`, and reproduce Figure 2.1.
  b. Show that the triplets `randu[i,]` lie on one of 15 parallel hyperplanes.

Exercise 2.11 In both questions, the comparison between generators is understood in terms of efficiency via the `system.time` function.

  a. Generate a binomial $Bin(n, p)$ random variable with $n = 25$ and $p = .2$. Plot a histogram for a simulated sample and compare it with the binomial mass function. Compare your generator with the R binomial generator.
  b. For $\alpha \in [0, 1]$, show that the R code

```
> u=runif(1)
> while(u > alpha) u=runif(1)
> U=u
```

  produces a random variable U from $\mathcal{U}([0, \alpha])$. Compare it with the transform αU, $U \sim \mathcal{U}(0, 1)$, for values of α close to 0 and close to 1, and with `runif(1,max=alpha)`.

Exercise 2.12 Referring to Example 2.2,

  a. Generate gamma and beta random variables according to (2.1).
  b. Show that if $U \sim \mathcal{U}_{[0,1]}$, then $X = -\log U/\lambda \sim \mathcal{E}xp(\lambda)$.
  c. Show that if $U \sim \mathcal{U}_{[0,1]}$, then $X = \log \frac{u}{1-u}$ is a Logistic(0, 1) random variable.

Exercise 2.13 The Pareto $\mathcal{P}(\alpha)$ distribution is defined by its density $f(x|\alpha) = \alpha x^{-\alpha-1}$ over $(1, \infty)$. Show that it can be generated as the $-1/\alpha$ power of a uniform variate. Plot the histogram and the density.

Exercise 2.14 Referring to Example 2.5:

  a. Verify the R code for the Poisson generator. Compare it with `rpois`.

b. The *negative binomial distribution*, with parameters r and p, has mass function

$$P(Y = y) = \binom{r + y - 1}{y} p^r (1 - p)^y, \quad y = 0, 1, \ldots,$$

with mean $r(1 - p)/p$ and variance $r(1 - p)/p^2$. For $r = 10$ and $p = .01, .1, .5$, generate 1000 random variables and draw their histograms. Compare the histograms with the probability functions and your generator with rnegbin (which is in the MASS package).

c. The *logarithmic series* distribution has mass function

$$P(X = x) = \frac{-(1 - p)^x}{x \log p}, \quad x = 1, 2, \ldots, \quad 0 < p < 1.$$

For $p = .001, .01, .5$, generate 1000 random variables and draw a histogram. Compare the histograms with the probability functions.

Exercise 2.15 The Poisson distribution $\mathcal{P}(\lambda)$ is connected to the exponential distribution through the Poisson process in that it can be simulated by generating exponential random variables until their sum exceeds 1. That is, if $X_i \sim \mathcal{E}xp(\lambda)$ and if K is the first value for which $\sum_{i=1}^{K+1} X_i > 1$, then $K \sim \mathcal{P}(\lambda)$. Compare this algorithm with rpois and the algorithm of Example 2.5 for both small and large values of λ.

Exercise 2.16 An algorithm to generate beta random variables was given in Example 2.2 for $\alpha \geq 1$ and $\beta \geq 1$. Another algorithm is based on the following property: If U and V are iid $\mathcal{U}_{[0,1]}$, the distribution of

$$\frac{U^{1/\alpha}}{U^{1/\alpha} + V^{1/\beta}},$$

conditional on $U^{1/\alpha} + V^{1/\beta} \leq 1$, is the $\mathcal{B}e(\alpha, \beta)$ distribution. Compare this algorithm with rbeta and the algorithm of Example 2.2 for both small and large values of α, β.

Exercise 2.17 We saw in Example 2.2 that, if $\alpha \in \mathbb{N}$, the gamma distribution $\mathcal{G}a(\alpha, \beta)$ can be represented as the sum of α exponential random variables $\epsilon_i \sim \mathcal{E}xp(\beta)$. When $\alpha \notin \mathbb{N}$, this representation does not hold.

a. Show that we can assume $\beta = 1$ by using the transformation $y = \beta x$.
b. When the $\mathcal{G}(n, 1)$ distribution is generated from an $\mathcal{E}xp(\lambda)$ distribution, determine the optimal value of λ.
c. When $\alpha \geq 1$, show that we can use the Accept–Reject algorithm with candidate distribution $\mathcal{G}a(a, b)$ to generate a $\mathcal{G}a(\alpha, 1)$ distribution, as long as $a \leq \alpha$. Show that the ratio f/g is $b^{-a} x^{\alpha - a} \exp\{-(1 - b)x\}$, up to a normalizing constant, yielding the bound

$$M = b^{-a} \left(\frac{\alpha - a}{(1 - b)e} \right)^{\alpha - a}$$

for $b < 1$.
d. Show that the maximum of $b^{-a}(1 - b)^{\alpha - a}$ is attained at $b = a/\alpha$, and hence the optimal choice of b for simulating $\mathcal{G}a(\alpha, 1)$ is $b = a/\alpha$, which gives the same mean for $\mathcal{G}a(\alpha, 1)$ and $\mathcal{G}a(\alpha, b)$.
e. Defend the choice of $a = \lfloor \alpha \rfloor$ as the best choice of a among the integers.
f. Discuss the strategy to adopt when $\alpha < 1$.

Exercise 2.18 The rather strange density

$$f(x) \propto \exp(-x^2/2) \left\{ \sin(6x)^2 + 3\cos(x)^2 \sin(4x)^2 + 1 \right\}$$

can be generated using the Accept–Reject algorithm.

a. Plot $f(x)$ and show that it can be bounded by $Mg(x)$, where g is the standard normal density $g(x) = \exp(-x^2/2)/\sqrt{2\pi}$. Find an acceptable if not necessarily optimal value of M. (*Hint:* Use the function `optimise`.)

b. Generate 2500 random variables from f using the Accept–Reject algorithm.

c. Deduce from the acceptance rate of this algorithm an approximation of the normalizing constant of f, and compare the histogram with the plot of the normalized f.

Exercise 2.19 In an Accept–Reject algorithm that generates a $\mathcal{N}(0,1)$ random variable from a double-exponential distribution with density $g(x|\alpha) = (\alpha/2)\exp(-\alpha|x|)$, compute the upper bound M over f/g and show that the choice $\alpha = 1$ optimizes the corresponding acceptance rate.

Exercise 2.20 In each of the following cases, construct an Accept–Reject algorithm, generate a sample of the corresponding random variables, and draw the density function on top of the histogram.

a. Generate normal random variables using a Cauchy candidate in Accept–Reject.

b. Generate gamma $\mathcal{G}(4.3, 6.2)$ random variables using a gamma $\mathcal{G}(4,7)$ candidate.

Exercise 2.21 The noncentral chi-squared distribution, $\chi_p^2(\lambda)$, can be defined by

(i). a mixture representation (2.2), where $g(x|y)$ is the density of χ_{p+2y}^2 and $p(y)$ is the density of $\mathcal{P}(\lambda/2)$, and

(ii). the sum of a χ_{p-1}^2 random variable and the square of a $\mathcal{N}(\|\theta\|, 1)$.

a. Show that both those representations hold.

b. Show that the representations are equivalent if $\lambda = \theta^2/2$.

c. Compare the corresponding algorithms that can be derived from these representations among themselves and also with `rchisq` for small and large values of λ.

Exercise 2.22 *Truncated normal distributions* $\mathcal{N}^+(\mu, \sigma^2, a)$, in which the range of a normal random variable is truncated, appear in many contexts. If $X \sim \mathcal{N}(\mu, \sigma^2)$, conditional on the event $\{x \geq a\}$, its density is proportional to

$$\exp\left\{ -(x-\mu)^2/2\sigma^2 \right\} \, \mathbb{I}_{x \geq a}.$$

a. The naïve method of simulating this random variable is to generate a $\mathcal{N}(\mu, \sigma^2)$ until the generated value is larger than a. Implement the R code

```
> Nsim=10^4
> X=rep(0,Nsim)
> for (i in 1:Nsim){
+   z=rnorm(1,mean=mu,sd=sigma)
+   while(z<a) z=rnorm(1,mean=mu,sd=sigma)
+   X[i]=z}
```

and evaluate the algorithm for $\mu = 0$, $\sigma = 1$, and various values of a.

b. Show that the algorithm in part a requires, on average, $1/\Phi((\mu-a)/\sigma)$ simulations from $\mathcal{N}(\mu,\sigma^2)$ for one acceptance. Deduce that, if a is in the tail of the distribution, this algorithm will take a very long time.

c. We now consider the case where $\mu = 0$ and $\sigma = 1$. Show that an Accept–Reject algorithm based on a normal $\mathcal{N}(\overline{\mu},1)$ candidate can be implemented to generate from the $\mathcal{N}^+(0,1,a)$ distribution for $\overline{\mu} > 0$. For a given a, discuss the optimization in $\overline{\mu}$.

d. Another potential candidate distribution for the Accept–Reject algorithm is the translated exponential distribution, $\mathcal{E}xp(\alpha,a)$, with density

$$g_\alpha(z) = \alpha e^{-\alpha(z-a)}\, \mathbb{I}_{z\ge a}\,.$$

Show that the ratio $(f/g_\alpha)(z) \propto e^{\alpha(z-a)}\, e^{-z^2/2}$ is then bounded by $\exp(\alpha^2/2 - \alpha a)$ if $\alpha \ge a$. Deduce that $a = \alpha$ gives a legitimate candidate density. Compare the performance of the corresponding Accept–Reject algorithm based on an $\mathcal{E}xp(a,a)$ candidate with the algorithm in part c, especially for a located in the tails of the normal $\mathcal{N}(0,1)$ distribution.

(The scale of the exponential distribution in part d can be optimized, but this may not lead to explicit expressions for the candidate scale. Two-sided normal truncation (that is, when $b \le x \le a$) is a bit more tricky to deal with. See Robert (1995b) for a resolution, or use rtrun from the package bayesm. Also see Exercise 7.21 for another truncated normal generator.)

Exercise 2.23 Given a sampling density $f(x|\theta)$ and a prior density $\pi(\theta)$, if we observe $\mathbf{x} = x_1,\ldots,x_n$, the posterior distribution of θ is

$$\pi(\theta|\mathbf{x}) = \pi(\theta|x_1,\ldots,x_n) \propto \prod_i f(x_i|\theta)\pi(\theta),$$

where $\prod_i f(x_i|\theta) = L(\theta|x_1,\ldots,x_n)$ is the likelihood function.

a. If $\pi(\theta|\mathbf{x})$ is the target density in an Accept–Reject algorithm, and if $\pi(\theta)$ is the candidate density, show that the optimal bound M is the likelihood function evaluated at the MLE.

b. For estimating a normal mean, a robust prior is the Cauchy. For $X_i \sim \mathcal{N}(\theta,1)$, $\theta \sim \mathcal{C}(0,1)$, the posterior distribution is

$$\pi(\theta|\mathbf{x}) \propto \frac{1}{\pi}\frac{1}{1+\theta^2}\frac{1}{2\pi}\prod_{i=1}^{n} e^{-(x_i-\theta)^2/2}.$$

Set $\theta_0 = 3$, $n = 10$, and generate $X_1,\ldots,X_n \sim \mathcal{N}(\theta_0,1)$. Use the Accept–Reject algorithm with a Cauchy $\mathcal{C}(0,1)$ candidate to generate a sample from the posterior distribution. Evaluate how well the value θ_0 is recovered. How much better do things get if n is increased?

3

Monte Carlo Integration

*"Every time I think I know what's going on, suddenly there's another
layer of complications. I just want this damn thing solved."*

John Scalzi
The Last Colony

Reader's guide

While Chapter 2 focused on the simulation techniques useful to produce random
variables by computer, this chapter introduces the major concepts of Monte Carlo
methods; that is, taking advantage of the availability of computer-generated random variables to approximate univariate and multidimensional integrals. In Section
3.2, we introduce the basic notion of Monte Carlo approximations as a by-product
of the Law of Large Numbers, while Section 3.3 highlights the universality of
the approach by stressing the versatility of the representation of an integral as
an expectation. Chapter 5 will similarly deal with the resolution of optimization
problems by simulation techniques.

C.P. Robert, G. Casella, *Introducing Monte Carlo Methods with R*, Use R,
DOI 10.1007/978-1-4419-1576-4_3, © Springer Science+Business Media, LLC 2010

3.1 Introduction

Two major classes of numerical problems that arise in statistical inference are *optimization* problems and *integration* problems. Indeed, numerous examples (see Rubinstein, 1981, Gentle, 2002, or Robert, 2001) show that it is not always possible to analytically compute the estimators associated with a given paradigm (maximum likelihood, Bayes, method of moments, etc.).

Thus, whatever the type of statistical inference, we are often led to consider numerical solutions. The previous chapter introduced a number of methods for the computer generation of random variables with any given distribution and hence provides a basis for the construction of solutions to our statistical problems. A general solution is indeed to use simulation, of either the true or some substitute distributions, to calculate the quantities of interest. In the setup of decision theory, whether it is classical or Bayesian, this solution is natural since risks and Bayes estimators involve integrals with respect to probability distributions.

Note that the possibility of producing an almost infinite number of random variables distributed according to a given distribution gives us access to the use of *frequentist* and *asymptotic* results much more easily than in the usual inferential settings, where the sample size is most often fixed. One can therefore apply probabilistic results such as the Law of Large Numbers or the Central Limit Theorem, since they allow assessment of the convergence of simulation methods (which is equivalent to the deterministic bounds used by numerical approaches).

Before embarking upon the description of Monte Carlo techniques, note that an apparently obvious alternative to the use of simulation methods for approximating integrals of the form

$$\int_{\mathcal{X}} h(x) \, f(x) \, \mathrm{d}x,$$

where f is a probability density, would be to rely on numerical methods such as Simpson's and the trapezium rules. For instance, R offers two related functions that run unidimensional integration, `area` (in the `MASS` library) and `integrate`. However, `area` cannot deal with infinite bounds in the integral and therefore requires some prior knowledge of the region of integration. The other function, `integrate`, accepts infinite bounds but is unfortunately very fragile and can produce untrustworthy output.

Example 3.1. As a test, we compare the use of `integrate` on the integral

$$\int_0^\infty x^{\lambda-1} \exp(-x) \, \mathrm{d}x$$

with the computation of $\Gamma(\lambda)$ via the `gamma` function. Implementing this comparison as

```
> ch=function(la){
+    integrate(function(x){x^(la-1)*exp(-x)},0,Inf)$val}
> plot(lgamma(seq(.01,10,le=100)),log(apply(as.matrix(
+ seq(.01,10,le=100)),1,ch)),xlab="log(integrate(f))",
+ ylab=expression(log(Gamma(lambda))),pch=19,cex=.6)
```

we obtain the sequence represented in Figure 3.1, which does not show any discrepancy even for very small values of λ.   ◄

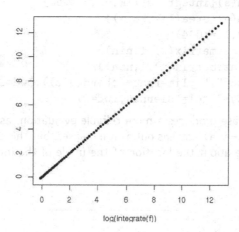

log(integrate(f))

Fig. 3.1. Comparison of the integrate evaluation of the $\Gamma(\lambda)$ integral with its true value.

A main difficulty with numerical integration methods such as integrate is that they often fail to spot the region of importance for the function to be integrated. In contrast, simulation methods naturally target this region by exploiting the information provided by the probability density associated with the integrals.

Example 3.2. Consider a sample of ten Cauchy rv's x_i ($1 \leq i \leq 10$) with location parameter $\theta = 350$. The (pseudo-) marginal of the sample under a flat prior is then

$$m(\mathbf{x}) = \int_{-\infty}^{\infty} \prod_{i=1}^{10} \frac{1}{\pi} \frac{1}{1 + (x_i - \theta)^2} \, d\theta.$$

However, integrate returns a wrong numerical value

```
> cac=rcauchy(10)+350
> lik=function(the){
+    u=dcauchy(cac[1]-the)
+    for (i in 2:10)
```

```
+        u=u*dcauchy(cac[i]-the)
+    return(u)}
>  integrate(lik,-Inf,Inf)
7.38034e-46 with absolute error < 1.5e-45
>  integrate(lik,200,400)
4.83155e-13 with absolute error < 9e-13
```

and fails to signal the difficulty since the error evaluation is absurdly small. Furthermore, the result is not comparable to area:

```
> cac=rcauchy(10)
> nin=function(a){integrate(lik,-a,a)$val}
> nan=function(a){area(lik,-a,a)}
> x=seq(1,10^3,le=10^4)
> y=log(apply(as.matrix(x),1,nin))
> z=log(apply(as.matrix(x),1,nan))
> plot(x,y,type="l",ylim=range(cbind(y,z)),lwd=2)
> lines(x,z,lty=2,col="sienna",lwd=2)
```

Using area in that case produces a more reliable evaluation, as shown in Figure 3.2, since area(lik,-a,a) flattens out as a increases, but this obviously requires some prior knowledge about the location of the mode of the integrand.   ◄

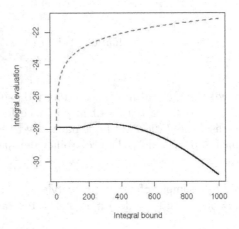

Fig. 3.2. Comparison of integrate and area on the integral of a Cauchy likelihood in log scale (the outcome of area corresponds to the dashed curve above).

Lastly, numerical integration tools cannot easily face the highly (or even moderately) multidimensional integrals that are the rule in statistical problems. Devising specific integration tools for those problems would be too costly, especially because we can take advantage of the probabilistic nature of those integrals.

3.2 Classical Monte Carlo integration

Before applying our simulation techniques to practical problems, let us recall
the properties that justify their use, referring to Robert and Casella (2004)
for (many) more details. The generic problem is about evaluating the integral

$$(3.1) \qquad \mathbb{E}_f[h(X)] = \int_{\mathcal{X}} h(x)\, f(x)\, \mathrm{d}x,$$

where $\mathcal{X}$ denotes the set where the random variable X takes its values, which
is usually equal to the support of the density f. The principle of the Monte
Carlo method for approximating (3.1) is to generate a sample $(X_1, \ldots, X_n)$
from the density f and propose as an approximation the empirical average

$$\overline{h}_n = \frac{1}{n} \sum_{j=1}^{n} h(x_j),$$

computed by `mean(h(x))` in R, since $\overline{h}_n$ converges almost surely (i.e. for almost
every generated sequence) to $\mathbb{E}_f[h(X)]$ by the Strong Law of Large Numbers.
Moreover, when $h^2(X)$ has a finite expectation under f, the speed of convergence of $\overline{h}_n$ can be assessed since the convergence takes place at a speed
$O(\sqrt{n})$ and the asymptotic variance of the approximation is

$$\mathrm{var}(\overline{h}_n) = \frac{1}{n} \int_{\mathcal{X}} (h(x) - \mathbb{E}_f[h(X)])^2 f(x) \mathrm{d}x,$$

which can also be estimated from the sample $(X_1, \ldots, X_n)$ through

$$v_n = \frac{1}{n^2} \sum_{j=1}^{n} [h(x_j) - \overline{h}_n]^2.$$

More specifically, due to the Central Limit Theorem, for large n,

$$\frac{\overline{h}_n - \mathbb{E}_f[h(X)]}{\sqrt{v_n}}$$

is approximately distributed as a $\mathcal{N}(0,1)$ variable, and this leads to the construction of a convergence test and confidence bounds on the approximation
of $\mathbb{E}_f[h(X)]$.

Example 3.3. For the toy function[1]

$$(3.2) \qquad h(x) = [\cos(50x) + \sin(20x)]^2,$$

represented in the upper panel of Figure 3.3, consider evaluating its integral
over $[0,1]$. It can be seen as a uniform expectation, and therefore we gener-
ate $U_1, U_2, \ldots, U_n$ iid $\mathcal{U}(0,1)$ random variables and approximate $\int h(x)\mathrm{d}x$ with

---

[1] This function can be integrated analytically.

$\sum h(U_i)/n$. The lower panel in Figure 3.3 shows the running means and the bounds derived from the estimated standard errors against the number n of simulations. The R implementation is as follows:

```
> h=function(x){(cos(50*x)+sin(20*x))^2}
> par(mar=c(2,2,2,1),mfrow=c(2,1))
> curve(h,xlab="Function",ylab="",lwd=2)
> integrate(h,0,1)
0.965201 with absolute error < 1.9e-10
> x=h(runif(10^4))
> estint=cumsum(x)/(1:10^4)
> esterr=sqrt(cumsum((x-estint)^2))/(1:10^4)
> plot(estint, xlab="Mean and error range",type="l",lwd=
+ 2,ylim=mean(x)+20*c(-esterr[10^4],esterr[10^4]),ylab="")
> lines(estint+2*esterr,col="gold",lwd=2)
> lines(estint-2*esterr,col="gold",lwd=2)
```

Note that the confidence band produced in this figure is not a 95% confidence band in the classical sense (i.e., it does not correspond to a confidence band on the graph of estimates, but rather to the confidence assessment that you can produce for every number of iterations, were you to stop at this number of iterations). ◄

⚡ While the bonus brought by the simultaneous evaluation of the error of the Monte Carlo estimate cannot be disputed, you must be aware that it is only trustworthy as far as v_n is a proper estimate of the variance of $\overline{h}_n$. In critical situations where v_n does not converge at all or does not even converge fast enough for a CLT to apply, this estimate and the confidence region associated with it cannot be trusted.

When monitoring Monte Carlo convergence, an issue that will be fully addressed in the next chapter, the R command cumsum is quite handy in that it computes all the partial sums of a sequence at once and thus allows the immediate representation of the sequence of estimators.

Exercise 3.1 For the normal-Cauchy Bayes estimator

$$\delta(x) = \int_{-\infty}^{\infty} \frac{\theta}{1+\theta^2} e^{-(x-\theta)^2/2} \, d\theta \Big/ \int_{-\infty}^{\infty} \frac{1}{1+\theta^2} e^{-(x-\theta)^2/2} \, d\theta,$$

solve the following questions when $x = 0, 2, 4$.

a. Plot the integrands, and use Monte Carlo integration based on a Cauchy simulation to calculate the integrals.
b. Monitor the convergence with the standard error of the estimate. Obtain three digits of accuracy with probability .95.
c. Repeat the experiment with a Monte Carlo integration based on a normal simulation and compare both approaches.

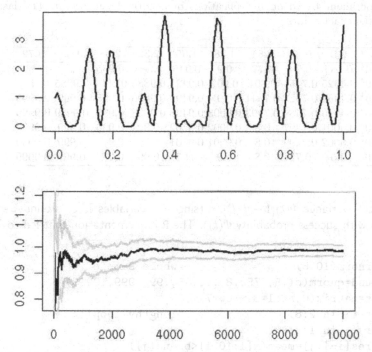

Fig. 3.3. Approximation of the integral of the function (3.2): *(upper)* function (3.2), and *(lower)* mean ± two standard errors against iterations for a single sequence of simulations.

The Monte Carlo methodology illustrated by the example above can be successfully implemented in a wide range of cases where the distributions involved in the model can be simulated. For instance, we could use Monte Carlo sums to calculate a normal cumulative distribution function (even though the normal cdf can now be found in all software and many pocket calculators).

Example 3.4. Given a normal $\mathcal{N}(0,1)$ sample of size n, $(x_1, \ldots, x_n)$, the approximation of

$$\Phi(t) = \int_{-\infty}^{t} \frac{1}{\sqrt{2\pi}} e^{-y^2/2} dy$$

by the Monte Carlo method is

$$\hat{\Phi}(t) = \frac{1}{n} \sum_{i=1}^{n} \mathbb{I}_{x_i \leq t},$$

Table 3.1. Evaluation of some normal probabilities $\Pr(X \leq t)$ by a regular Monte Carlo experiment based on n replications of a normal generation. The last line achieves the exact values.

n/t	0.0	0.67	0.84	1.28	1.65	2.32	2.58	3.09	3.72
10^2	0.485	0.74	0.77	0.9	0.945	0.985	0.995	1	1
10^3	0.4925	0.7455	0.801	0.902	0.9425	0.9885	0.9955	0.9985	1
10^4	0.4962	0.7425	0.7941	0.9	0.9498	0.9896	0.995	0.999	0.9999
10^5	0.4995	0.7489	0.7993	0.9003	0.9498	0.9898	0.995	0.9989	0.9999
10^6	0.5001	0.7497	0.8	0.9002	0.9502	0.99	0.995	0.999	0.9999
10^7	0.5002	0.7499	0.8	0.9001	0.9501	0.99	0.995	0.999	0.9999
10^8	0.5	0.75	0.8	0.9	0.95	0.99	0.995	0.999	0.9999

with (exact) variance $\Phi(t)[1 - \Phi(t)]/n$ (since the variables $\mathbb{I}_{x_i \leq t}$ are independent Bernoulli with success probability $\Phi(t)$). The R implementation that led to Table 3.1 is

```
> x=rnorm(10^8)                    #whole sample
> bound=qnorm(c(.5,.75,.8,.9,.95,.99,.999,.9999))
> res=matrix(0,ncol=8,nrow=7)
> for (i in 2:8)                   #lengthy loop!!
+ for (j in 1:8)
+   res[i-1,j]=mean(x[1:10^i]<bound[j])
> matrix(as.numeric(format(res,digi=4)),ncol=8)
```

For values of t around $t = 0$, the variance is thus approximately $1/4n$, and to achieve a precision of four decimals, we need $2 \times \sqrt{1/4n} \leq 10^{-4}$ simulations, i.e., about $n = (10^4)^2 = 10^8$ simulations. Table 3.1 gives the evolution of this approximation for several values of t and shows an accurate evaluation for 100 million iterations. Note that greater (absolute) accuracy is achieved in the tails and that (much) more efficient simulation methods could be used.   ◀

As you have presumably noticed, the outputs in R are represented with all the available digits, as in

```
> rnorm(1)
[1] -0.08581098
```

While this is logical from an informatic point of view, it is not recommended to produce all those digits in statistical and simulation environments because most of them are not significant and also because it impairs the readability of the output. The `format` function is then quite handy to cut down the number of represented digits, as shown in the last line of the R program above.

The Monte Carlo approximation of a probability distribution function illustrated by Example 3.4 has nontrivial applications since it can be used in assessing the distribution of a test statistic such as a likelihood ratio test under a null hypothesis, as illustrated in Robert and Casella (2004), as well as its power under alternatives.

It may thus seem at this stage that the Monte Carlo methodology introduced in this section is sufficient to approximate integrals like (3.1) in a controlled way. However, while the straightforward Monte Carlo method indeed provides good approximations of (3.1) in most regular cases, there exist more efficient alternatives that not only avoid a direct simulation from f but also can be used repeatedly for several integrals of the form (3.1). The repeated use can be for either a family of functions h or a family of densities f. In addition, problems of tail simulation as in Example 3.4 can be processed much more efficiently than simulating from f since simulating events with a very small probability requires a very large number of simulations under f to achieve a given (relative) precision.

Exercise 3.2 Given that $\mathbb{I}_{X_i \leq t}$ is a Bernoulli random variable equal to 1 with probability $\Phi(t)$, show that the variance of the normalized estimator $\mathbb{I}_{X_i \leq t}/\Phi(t)$ goes to infinity when t decreases to $-\infty$. Deduce the number of simulations (as a function of t) that are necessary to achieve a variance less than 10^{-8}.

Exercise 3.3 If we are interested in the tail probability $\Pr(X > 20)$ when $X \sim \mathcal{N}(0,1)$, simulating from a $\mathcal{N}(0,1)$ distribution does not work. Express the probability as an integral and use an obvious change of variable to rewrite this integral as an expectation under a $\mathcal{U}(0,1/20)$ distribution. Deduce a Monte Carlo approximation to $\Pr(X > 20)$ along with an error assessment.

3.3 Importance sampling

The method we now study is called *importance sampling* because it relies on so-called *importance functions*, which are instrumental distributions , in lieu of the original distributions. In fact, an evaluation of (3.1) based on simulations from f is almost never optimal in the sense that using alternative distributions can improve the variance of the resulting estimator of (3.1).

3.3.1 An arbitrary change of reference measure

The importance sampling method is based on an alternative representation of (3.1). Given an arbitrary density g that is strictly positive when $h \times f$ is different from zero, we can indeed rewrite (3.1) as

$$(3.3) \qquad \mathbb{E}_f[h(X)] = \int_{\mathcal{X}} h(x) \frac{f(x)}{g(x)} g(x) \, dx = \mathbb{E}_g\left[\frac{h(X)f(X)}{g(X)}\right];$$

that is, as an expectation under the density g. (Note that $\mathcal{X}$ is again the set where X takes its value and that it may therefore be *smaller* than the support of the density g.) This *importance sampling fundamental identity* justifies the use of the estimator

$$(3.4) \qquad \frac{1}{n} \sum_{j=1}^{n} \frac{f(X_j)}{g(X_j)} h(X_j) \rightarrow \mathbb{E}_f[h(X)]$$

based on a sample $X_1, \dots, X_n$ generated from g (not from f!). Indeed, because (3.1) can thus be written as an expectation under g, (3.4) does converge to (3.1) for the same reason the regular Monte Carlo estimator $\overline{h}_n$ converges, whatever the choice of the distribution g (as long as $\mathrm{supp}(g) \supset \mathrm{supp}(h \times f)$). This ubiquitous property relates to the fact that (3.1) can be represented in an infinite number of ways by pairs (h, f) and thus that a given integral is not intrinsically associated with a given distribution. On the contrary, there is almost absolute freedom in its representation as an expectation.

⚡ The constraint on the support of g, $\mathrm{supp}(g) \supset \mathrm{supp}(h \times f)$, is absolute in that using a smaller support truncates the integral (3.3) and thus produces a biased result. This means, in particular, that when considering non-parametric solutions for g, the support of the kernel must be unrestricted.

Exercise 3.4 For the computation of the expectation $\mathbb{E}_f[h(X)]$ when f is the normal pdf and $h(x) = \exp(-(x-3)^2/2) + \exp(-(x-6)^2/2)$:

  a. Show that $\mathbb{E}_f[h(X)]$ can be computed in closed form and derive its value.
  b. Construct a regular Monte Carlo approximation based on a normal $\mathcal{N}(0,1)$ sample of size Nsim=10^3 and produce an error evaluation.
  c. Compare the above with an importance sampling approximation based on an importance function g corresponding to the $\mathcal{U}(-8,-1)$ distribution and a sample of size Nsim=10^3. (Warning: This choice of g does not provide a converging approximation of $\mathbb{E}_f[h(X)]$!)

Example 3.5. As mentioned at the end of Example 3.4, approximating tail probabilities using standard Monte Carlo sums breaks down once one goes far enough into the tails. For example, if $Z \sim \mathcal{N}(0,1)$ and we are interested in the probability $P(Z > 4.5)$, which is very small,

```
> pnorm(-4.5,log=T)
[1] -12.59242
```

simulating $Z^{(i)} \sim \mathcal{N}(0,1)$ only produces a hit once in about 3 million iterations! Of course, the problem is that we are now interested in the probability of a very rare event and thus naïve simulation from f will require a huge number of simulations to get a stable answer. However, thanks to importance sampling, we can greatly improve our accuracy and thus bring down the number of simulations by several orders of magnitude.

For instance, if we consider a distribution with support restricted to $(4.5, \infty)$, the additional and unnecessary variation of the Monte Carlo estimator due to simulating zeros (i.e., when $x < 4.5$) disappears. A natural choice is to take g as the density of the exponential distribution $\mathcal{E}xp(1)$ truncated at 4.5,

$$g(y) = e^{-y} \Big/ \int_{4.5}^{\infty} e^{-x}\mathrm{d}x = e^{-(y-4.5)},$$

and the corresponding importance sampling estimator of the tail probability is

$$\frac{1}{n}\sum_{i=1}^{n} \frac{f(Y^{(i)})}{g(Y^{(i)})} = \frac{1}{n}\sum_{i=1}^{n} \frac{e^{-Y_i^2/2+Y_i-4.5}}{\sqrt{2\pi}},$$

where the Y_i's are iid generations from g. The corresponding code is

```
> Nsim=10^3
> y=rexp(Nsim)+4.5
> weit=dnorm(y)/dexp(y-4.5)
> plot(cumsum(weit)/1:Nsim,type="l")
> abline(a=pnorm(-4.5),b=0,col="red")
```

The final value is then $3.312\,10^{-6}$, to be compared with the true value of 3.398×10^{-6}. As shown in Figure 3.4, the accuracy of the approximation is remarkable, especially when compared with the original size requirements imposed by a normal simulation. ◀

Exercise 3.5 In the exercise above, examine the impact of using a truncated exponential distribution $\mathcal{E}xp(\lambda)$ on the variance of the approximation of the tail probability.

Importance sampling is therefore of considerable interest since it puts very little restriction on the choice of the instrumental distribution g, which can be chosen from distributions that are either easy to simulate or efficient in the approximation of the integral. Moreover, the same sample (generated from g) can be used repeatedly, not only for different functions h but also for different densities f.

Example 3.6. This example stems from a Bayesian setting: When considering an observation x from a beta $\mathcal{B}(\alpha, \beta)$ distribution,

$$x \sim \frac{\Gamma(\alpha+\beta)}{\Gamma(\alpha)\Gamma(\beta)} x^{\alpha-1}(1-x)^{\beta-1}\,\mathbb{I}_{[0,1]}(x),$$

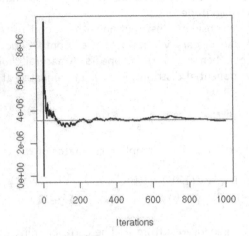

Fig. 3.4. Convergence of the importance sampling approximation of the normal tail probability $P(Z \geq 4.5)$, based on a sequence simulated from a translated exponential distribution. The straight line corresponds to the true value of the integral.

there exists a family of conjugate priors on (α, β) of the form

$$\pi(\alpha, \beta) \propto \left\{ \frac{\Gamma(\alpha + \beta)}{\Gamma(\alpha)\Gamma(\beta)} \right\}^{\lambda} x_0^{\alpha} y_0^{\beta},$$

where λ, x_0, y_0 are hyperparameters, since the posterior is then equal to

$$\pi(\alpha, \beta | x) \propto \left\{ \frac{\Gamma(\alpha + \beta)}{\Gamma(\alpha)\Gamma(\beta)} \right\}^{\lambda+1} [xx_0]^{\alpha} [(1 - x)y_0]^{\beta}.$$

This family of distributions is intractable if only because of the difficulty of dealing with gamma functions. Simulating directly from $\pi(\alpha, \beta | x)$ is therefore impossible. We thus need to use a substitute distribution $g(\alpha, \beta)$, and we can get a preliminary idea by looking at an image representation of $\pi(\alpha, \beta | x)$. If we take $\lambda = 1$, $x_0 = y_0 = .5$, and $x = .6$, the R code is

```
> f=function(a,b){
+    exp(2*(lgamma(a+b)-lgamma(a)-lgamma(b))+
+        a*log(.3)+b*log(.2))}
> aa=1:150      #alpha grid for image
> bb=1:100      #beta grid for image
> post=outer(aa,bb,f)
> image(aa,bb,post,xlab=expression(alpha),ylab=" ")
> contour(aa,bb,post,add=T)
```

The outer command is a handy abbreviation to compute a matrix A=outer(a,b,f) of dimension c(dim(a),dim(b)) whose A[i,j] element is equal to f(a[i],b[j]). While it is much faster than the basic double allocation loop,

```
> system.time(outer(aa,bb,f))
   user   system elapsed
  0.028    0.000   0.029
> system.time(for (j in 1:100){for (i in 1:150)
+ post[i,j]=f(a=aa[i],b=bb[j])})
   user   system elapsed
  0.360    0.004   0.367
```

it compares speedwise with a single allocation loop

```
> system.time(outer(aa,bb,f))
   user   system elapsed
  0.028    0.000   0.028
> system.time(for (j in 1:100){post[,j]=f(a=aa,b=bb[j])})
   user   system elapsed
  0.028    0.000   0.027
> system.time(for (i in 1:150){post[i,]=f(a=aa[i],b=bb)})
   user   system elapsed
  0.032    0.000   0.031
```

and thus does not offer a superefficient way to allocate values to a matrix.

The examination of Figure 3.5 *(left)* shows that a normal or a Student's t distribution on the pair (α, β) could be appropriate. Choosing a Student's $T(3, \mu, \Sigma)$ distribution with $\mu = (50, 45)$ and

$$\Sigma = \begin{pmatrix} 220 & 190 \\ 190 & 180 \end{pmatrix}$$

does produce a reasonable fit, as shown on Figure 3.5 *(right)* using the superposition of simulation from this $T(3, \mu, \Sigma)$ distribution with the surface of the posterior distribution. The covariance matrix above was obtained by trial-and-error, modifying the entries until the sample in Figure 3.5 *(right)* fit well enough:

```
> x=matrix(rt(2*10^4,3),ncol=2)        #T sample
> E=matrix(c(220,190,190,180),ncol=2)  #Scale matrix
> image(aa,bb,post,xlab=expression(alpha),ylab=" ")
> y=t(t(chol(E))%*%t(x)+c(50,45))
> points(y,cex=.6,pch=19)
```

Note the use of t(chol(E)) to ensure that the covariance matrix is E (up to a factor of 3 due to the use of the Student's t_3 distribution).

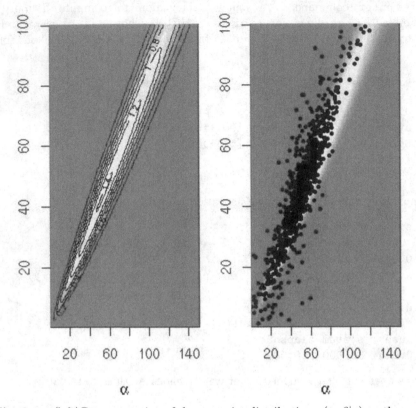

Fig. 3.5. *(left)* Representation of the posterior distribution $\pi(\alpha, \beta|x)$ on the parameters of a $\mathcal{B}(\alpha, \beta)$ distribution for $x = 0.6$. *(right)* Superposition of a sample of 10^3 points from a Student's t $\mathcal{T}(3, \mu, \Sigma)$ distribution used as an importance function.

If the quantity of interest is the marginal likelihood, as in Bayesian model comparison (Robert, 2001),

$$m(x) = \int_{\mathbb{R}_+^2} f(x|\alpha, \beta)\,\pi(\alpha, \beta)\,\mathrm{d}\alpha\mathrm{d}\beta$$

$$= \frac{\int_{\mathbb{R}_+^2} \left\{ \frac{\Gamma(\alpha+\beta)}{\Gamma(\alpha)\Gamma(\beta)} \right\}^{\lambda+1} [xx_0]^\alpha [(1-x)y_0]^\beta\,\mathrm{d}\alpha\mathrm{d}\beta}{x(1-x)\int_{\mathbb{R}_+^2} \left\{ \frac{\Gamma(\alpha+\beta)}{\Gamma(\alpha)\Gamma(\beta)} \right\}^\lambda x_0^\alpha y_0^\beta\,\mathrm{d}\alpha\mathrm{d}\beta},$$

we need to approximate both integrals and the same t sample can be used for both since the fit is equally reasonable on the prior surface. This approximation

$$\hat{m}(x) = \sum_{i=1}^{n} \left\{ \frac{\Gamma(\alpha_i + \beta_i)}{\Gamma(\alpha_i)\Gamma(\beta_i)} \right\}^{\lambda+1} [xx_0]^{\alpha_i} [(1-x)y_0]^{\beta_i} \Big/ g(\alpha_i, \beta_i) \Big/$$

(3.5) $$\sum_{i=1}^{n}\left\{\frac{\Gamma(\alpha_i+\beta_i)}{\Gamma(\alpha_i)\Gamma(\beta_i)}\right\}^{\lambda} x_0^{\alpha_i} y_0^{\beta_i}/g(\alpha_i,\beta_i),$$

where $(\alpha_i,\beta_i)_{1\leq i\leq n}$ are n iid realizations from g, is straightforward to implement in R:

```
> ine=apply(y,1,min)
> y=y[ine>0,]
> x=x[ine>0,]
> normx=sqrt(x[,1]^2+x[,2]^2)
> f=function(a) exp(2*(lgamma(a[,1]+a[,2])-lgamma(a[,1])
+    -lgamma(a[,2]))+a[,1]*log(.3)+a[,2]*log(.2))
> h=function(a) exp(1*(lgamma(a[,1]+a[,2])-lgamma(a[,1])
+    -lgamma(a[,2]))+a[,1]*log(.5)+a[,2]*log(.5))
> den=dt(normx,3)
> mean(f(y)/den)/mean(h(y)/den)
[1] 0.1361185
```

Our approximation of the marginal likelihood, based on those simulations is thus 0.1361. Similarly, the posterior expectations of the parameters α and β are obtained by

```
> mean(y[,1]*apply(y,1,f)/den)/mean(apply(y,1,h)/den)
[1] 19.33745
> mean(y[,2]*apply(y,1,f)/den)/mean(apply(y,1,h)/den)
[1] 16.54468
```

i.e., are approximately equal to 19.34 and 16.54, respectively. ◀

3.3.2 Sampling importance resampling

The importance sampling technique does more than approximate integrals, though, since it provides an alternative way to simulate from complex distributions. Recall that the method produces a sample $X_1,\ldots,X_n$ simulated from g along with its importance weights $f(X_i)/g(X_i)$. This sample can then be recycled by multinomial resampling into a sample that is (almost) from f.

Indeed, if we could sample with replacement from the weighted population $\{X_1,\ldots,X_n\}$, picking X_i with probability $f(X_i)/ng(X_i)$, we would get an outcome X^* distributed as

$$\Pr(X^*\in A)=\sum_{i=1}^{n}\Pr(X^*\in A \text{ and } X^*=X_i)$$

$$=\int_A \frac{f(x)}{g(x)} g(x)\,dx=\int_A f(x)\,dx,$$

and the method would then produce an exact simulation from f! Unfortunately, the probabilities $f(X_i)/ng(X_i)$ do not sum up to 1 (worse, some may even be larger than 1) and need to be renormalized into $(i=1,\ldots,n)$

(3.6) $$\omega_i = \frac{1}{n}\left\{f(X_i)/g(X_i)\right\} \Big/ \frac{1}{n}\sum_{j=1}^{n}\left\{f(X_j)/g(X_j)\right\}.$$

While the denominator is converging almost surely to one, the renormalization induces a bias in the distribution of the resampled values. Nonetheless, for large sample sizes, this bias is negligible, and we will thus use multinomial resampling (or an improved version; see Exercises 3.6 and 3.12) to approximate samples generated from f.

Exercise 3.6 Given an importance sample $(X_i, f(X_i)/g(X_i))$, show that if ω_i has a Poisson distribution $\omega_i \sim \mathcal{P}(f(X_i)/g(X_i))$, the estimator

$$\frac{1}{n}\sum_{i=1}^{n}\omega_i h(x_i)$$

is unbiased. Deduce that the sample derived by this sampling mechanism is marginally distributed from f.

The sole difficulty with the solution proposed in Exercise 3.6 is that the samples thus produced have a random size due to the random replications of each value in the weighted sample, ranging from 0 to ∞. While the setting where either f or g is missing a normalizing constant can be handled as well by replacing f/g by $\alpha f/g$, the impact on the final sample size is even harder to fathom (see Exercises 3.10 and 3.12).

The use of the renormalized weights in the importance sampling estimator produces the *self-normalized importance sampling estimator*

(3.7) $$\sum_{i=1}^{n} h(X_i)\,f(X_i)/g(X_i) \Big/ \sum_{j=1}^{n}\left\{f(X_j)/g(X_j)\right\},$$

which can also be used in situations when either f or g are missing a normalizing constant. The denominator of (3.6) is then estimating the missing constant(s) as well. (This is for instance the case in Example 3.6: The missing normalizing constant of the prior is estimated by `mean(apply(y,1,h)/den)` in the code above.)

⚡ The importance weights only provide a *relative* assessment of the adequacy of the simulated sample to the target density in that they indicate how much more likely X_i is to be simulated from f compared with X_j, but they should not be overinterpreted. For instance, if X_i has a self-normalized weight that is close to 1, it does not mean that this value is very likely to be generated from f but simply that it is much more likely than the other simulated values! Even when the fit between f and g is very poor, this occurrence is bound to happen. Therefore, more trustworthy indicators must be used to judge the adequacy of g against f.

Example 3.7. (Continuation of Example 3.6) The validity of the approximation (3.5) of the marginal likelihood, (i.e., the convergence of the importance sampling solution) can be assessed by graphical means as follows:

```
> par(mfrow=c(2,2),mar=c(4,4,2,1))
> weit=(apply(y,1,f)/den)/mean(apply(y,1,h)/den)
> image(aa,bb,post,xlab=expression(alpha),
+        ylab=expression(beta))
> points(y[sample(1:length(weit),10^3,rep=T,pro=weit),],
+ cex=.6,pch=19)
> boxplot(weit,ylab="importance weight")
> plot(cumsum(weit)/(1:length(weit)),type="l",
+      xlab="simulations", ylab="marginal likelihood")
> boot=matrix(0,ncol=length(weit),nrow=100)
> for (t in 1:100)
+    boot[t,]=cumsum(sample(weit))/(1:length(weit))
> uppa=apply(boot,2,quantile,.95)
> lowa=apply(boot,2,quantile,.05)
> polygon(c(1:length(weit),length(weit):1),c(uppa,rev(lowa)),
+         col="gold")
> lines(cumsum(weit)/(1:length(weit)),lwd=2)
> plot(cumsum(weit)^2/cumsum(weit^2),type="l",
+      xlab="simulations", ylab="Effective sample size",lwd=2)
```

We will not discuss in detail all those indicators, as some are explained in the next chapter. The upper left graph in Figure 3.6 shows that the sample weighted using the importance weight $\pi(\alpha_i, \beta_i|x)/g(\alpha_i, \beta_i)$ produces a fair rendering of a sample from $\pi(\alpha, \beta|x)$. The resampled points do not degenerate in a few points but instead cover, with high density, the correct range for the target distribution (compare it with the right-hand side of Figure 3.5). The upper right graph gives a representation of the spread of the importance weights. While there exist simulations with much higher weights than the others, the spread of the weight is not so extreme as to signify a degeneracy of the method. For instance, the highest reweighted point only represents 1% of the whole sample. The lower left graph represents the convergence of the estimator $\hat{m}(x)$ as n increases. The colored band surrounding the sequence is a bootstrap rendering (Section 1.5) of the variability of this estimator that mimics the confidence band represented in Figure 3.3 at a low cost. The lower right curve is representing the efficiency loss in using importance sampling by the effective sampling size (see Section 4.4),

$$\left\{ \sum_{i=1}^{n} \pi(\alpha_i, \beta_i|x)/g(\alpha_i, \beta_i) \right\}^2 \Big/ \sum_{i=1}^{n} \{\pi(\alpha_i, \beta_i|x)/g(\alpha_i, \beta_i)\}^2 \,,$$

which should be equal to n, if the (α_i, β_i)'s are generated from the posterior. The current graph shows that the sample produced has an efficiency of about 6%. We will further consider this indicator in Section 4.2.   ◀

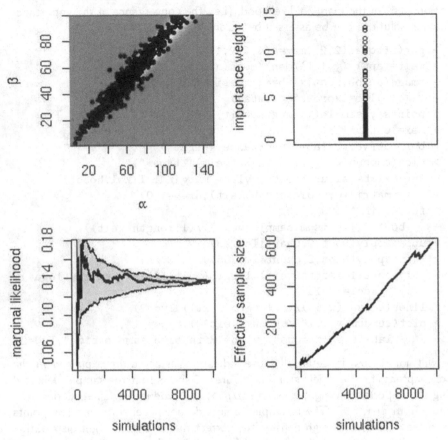

Fig. 3.6. *(upper left)* Superposition of 10^3 resampled points over the posterior distribution $\pi(\alpha, \beta|x)$ on the parameters of a $\mathcal{B}(\alpha, \beta)$ distribution for $x = 0.6$. *(upper right)* Boxplot of the importance weights. *(upper right)* Convergence of the approximation $\hat{m}(x)$ and bootstrap rendering of its variability. *(upper right)* Evolution of the effective sample size.

3.3.3 Selection of the importance function

The versatility of the importance sampling technique is high, but the downside of this versatility is that a poor choice of the importance function g may produce very poor outcomes. While the optimal choice of g is more of a theoretical exercise (see Rubinstein, 1981, or Robert and Casella, 2004, Theorem 3.12) than anything useful, an issue of primary relevance is to consider the variance of the resulting estimator (3.3) when judging the adequacy of the corresponding importance function g.

Indeed, while (3.4) does converge almost surely to (3.1), given that the expectation (3.1) exists, the variance of this estimator is finite only when the expectation

$$\mathbb{E}_g\left[h^2(X)\frac{f^2(X)}{g^2(X)}\right] = \mathbb{E}_f\left[h^2(X)\frac{f(X)}{g(X)}\right] = \int_{\mathcal{X}} h^2(x)\,\frac{f^2(x)}{g(x)}\,\mathrm{d}x < \infty$$

is finite. While not exactly prohibiting importance functions with tails lighter than those of f that lead to unbounded ratios f/g, this condition stresses that those functions are much more likely to lead to infinite variance estimators.

Before discussing this issue in more detail, let us consider a simple example to illustrate the disastrous impact of an infinite variance estimator.

Example 3.8. A simple setting where infinite variance occurs is when using a $\mathcal{N}(0,1)$ normal importance function aimed at a Cauchy $\mathcal{C}(0,1)$ target. The ratio $f(x)/g(x) \propto \exp(x^2/2)/(1+x^2)$ is then explosive in that even moderately high values of x get very large importance weights. If you run the code

```
> x=rnorm(10^6)
> wein=dcauchy(x)/dnorm(x)
> boxplot(wein/sum(wein))
> plot(cumsum(wein*(x>2)*(x<6))/cumsum(wein),type="l")
> abline(a=pcauchy(6)-pcauchy(2),b=0,col="sienna")
```

a few times, you should see graphs like the one in Figure 3.7 occurring, namely patterns with huge jumps in the cumulated average, even with a large number of terms in the average. The jumps happen at values of the simulations for which $\exp(x^2/2)/(1+x^2)$ is large, which means when x is large. The reason for this phenomenon is that since those values are rare under the normal importance distribution (meaning rarer than under the Cauchy target), they need to compensate for their rarity by taking high weights. For instance, in Figure 3.7, the major jump is due to a value of $x = 5.49$ associated with a normalized weight of $\omega_i = 0.094$. It means that this single point has a weight of about 10% in a sample of a million points! Obviously, it is impossible to trust the outcome of this simulation since the sample size is then irrelevant (i.e., most simulated values have a negligible weight). ◀

When the ratio f/g is unbounded, the importance weights $f(x_j)/g(x_j)$ often vary widely, giving too much importance to a few values x_j and thus degrading the efficiency of the estimator (3.4). As in the example above, it may happen that the estimate abruptly changes from one iteration to the next one, even after many iterations, due to a single simulation. Conversely, importance distributions g with thicker tails than f ensure that the behavior of the ratio f/g is not the cause of the divergence of $\mathbb{E}_f[h^2(X)f(X)/g(X)]$.

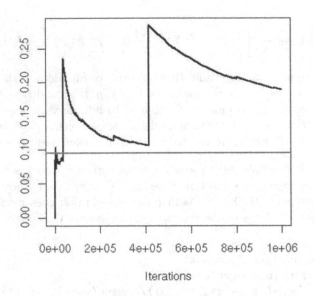

Fig. 3.7. Evolution of the importance sampling estimator of the probability $P(2 \leq Z \leq 6)$ against iteration indices, when Z is distributed from a Cauchy distribution and the importance function is normal. The straight line is the exact value, 0.095.

Using thicker-tailed importance sampling proposal distributions is almost a "must" when considering the approximation of functions h such that (3.1) exists but $\mathbb{E}_f[h^2(X)]$ does not. In such cases, using regular Monte Carlo is not possible, since the empirical average of the $h(X_i)$'s then has no variance.

Exercise 3.7 When f is a $\mathcal{T}_\nu$ distribution, show that the variance of the importance sampling estimator associated with an importance function g and the integrand $h(x) = \sqrt{x/(1-x)}$ is infinite for all g's such that $g(1) < \infty$. Discuss a sufficient condition on g for the variance to be finite. (*Hint:* See Example 3.9.)

⚡ The self-normalized estimator (3.7) requires the same condition as in the nonnormalized case for the variance to be finite. But, as detailed in Chapter 4, the expression of the variance is not available in closed form and needs to be approximated by Monte Carlo methods.

As a generic recommendation, at this stage we thus suggest looking for distributions g for which $|h|f/g$ is almost constant or at least enjoys a controlled tail behavior, since this is more likely to produce estimators with a finite variance.

A basic requirement for functions h with restricted supports as in Example 3.5 is that g adopt the same support as h unless this is prevented by the complexity of h. Obviously, this requires fitting a new importance function for each integrand h to be considered, but this is the price to pay to achieve (much) more efficiency, as shown by Example 3.5.

Given that importance sampling primarily applies in settings where f is not easy to study, this constraint on the tails of f is often not easy to implement, especially when the dimensionality is high. A generic solution nonetheless exists based on the artificial incorporation of a fat tail component in the importance function g. This solution is called *defensive sampling* by Hesterberg (1995) and can be achieved by substituting a mixture density for the density g,

$$(3.8) \qquad \rho g(x) + (1 - \rho)\ell(x), \qquad 0 < \rho < 1,$$

where ρ is close to 1 and the density ℓ is chosen for its heavy tails (for instance, a Cauchy or a Pareto distribution), not necessarily in conjunction with the problem at hand.

Assuming g is provided by the setting, choosing the heavy-tailed function ℓ is potentially delicate. In the special case of Bayesian inference, when the target distribution f is the posterior distribution, it is, however, natural to choose the prior if proper. Indeed, this function has heavier tails than f by construction and is usually a standard distribution that is easy to simulate. Using the prior as the main importance function g would not make sense because of the waste induced (assuming the data are informative). But using it as a stabilizing factor does make sense.

Exercise 3.8 (Smith and Gelfand, 1992) Show that when evaluating an integral based on a posterior distribution

$$\pi(\theta|x) \propto \pi(\theta)\ell(\theta|x),$$

where π is the prior distribution and ℓ the likelihood function, the prior distribution can always be used as an instrumental distribution.

a. Show that the variance of the weight is finite when the likelihood is bounded.
b. Compare the previous choice with choosing $\ell(\theta|x)$ as the instrumental distribution when the likelihood is proportional to a density. (*Hint:* Consider the case of exponential families.)
c. Discuss the drawbacks of this (these) choice(s) in specific settings.
d. Show that a mixture between both instrumental distributions can ease some of the drawbacks.

From an operational point of view, generating from (3.8) means that the observations are generated with probability ρ from g and with probability $1 - \rho$ from ℓ, using a code like

```
> mix=function(n=1,p=0.5){
+     m=rbinom(1,size,pro=p)
+     c(simg(m),siml(n-m))}
```

if `simg` and `siml` denote generators from g and ℓ, respectively. We stress that the fact that some points are generated from g and others from ℓ does not impact the importance weight in that it is equal to $f(x)/\{\rho g(x)+(1-\rho)\ell(x)\}$ for *all* generated values.

By construction, the importance sampling estimator integrates out the uniform variable used to decide between g and ℓ. Conditioning on this uniform variable would both induce more variability and destroy the purpose of using a mixture by dividing once again by $g(x)$ in the importance weight. (We discuss in detail such a marginalization perspective in the next chapter, in Section 4.6, where uniform variables involved in the simulation are integrated out in the estimator.)

Note that the selection of a random number of simulations from g and ℓ is *in fine* unnecessary, however, since generating exactly ρn x_i's from g and $(1 - \rho)n$ y_i's from ℓ produces an unbiased estimator (under the assumption that ρn is an integer) in the sense that the importance sampling estimator

$$\frac{1}{n} \sum_{i=1}^{\rho n} h(x_i)\frac{f(x_i)}{\rho g(x_i) + (1 - \rho)\ell(x_i)} + \frac{1}{n} \sum_{i=1}^{(1-\rho)n} h(y_i)\frac{f(y_i)}{\rho g(y_i) + (1 - \rho)\ell(y_i)}$$

has a global (if not termwise) expectation equal to $\mathbb{E}_f[h(X)]$ (see Owen and Zhou, 2000, for more details). Thus, simulating a fixed number of points from each distribution is both valid and interesting in that it completely eliminates the variability due to the binomial sampling above.

Example 3.9. As indicated in Exercise 3.7, the computation of the integral

$$(3.9) \qquad \int_1^\infty \sqrt{\frac{x}{x-1}}\, t_2(x)\, \mathrm{d}x = \int_1^\infty \sqrt{\frac{x}{x-1}} \frac{\Gamma(3/2)/\sqrt{2\pi}}{(1+x^2/2)^{3/2}}\, \mathrm{d}x$$

is delicate because the function $h(x) = \sqrt{1/(x-1)}$ is not square-integrable and therefore using simulations from the $\mathcal{T}_2$ distribution will produce an infinite variance for the Monte Carlo estimator of the integral.

This feature means that a mixture of the $\mathcal{T}_2$ density with a well-behaved ℓ is required. To achieve integrability of $h^2(x)f(x)/\ell(x)$ calls for ℓ to be divergent in $x = 1$ and for ℓ to decrease faster than x^5 when x goes to infinity. Those boundary conditions suggest that

$$\ell(x) \propto \frac{1}{\sqrt{x-1}} \frac{1}{x^{3/2}} \mathbb{I}_{x>1}$$

(which is defined up to a constant) is an acceptable density. To characterize this density, you can check that

$$\int_1^y \frac{dx}{\sqrt{x-1}x^{3/2}} = \int_0^{y-1} \frac{dw}{\sqrt{w}(w+1)^{3/2}}$$

$$= \int_0^{\sqrt{y-1}} \frac{2d\omega}{(\omega^2+1)^{3/2}}$$

$$= \int_0^{\sqrt{2(y-1)}} \frac{2dt}{(1+t^2/2)^{3/2}} \, .$$

This implies that $\ell(x)$ corresponds to the density of $(1 + T^2/2)$ when $T \sim \mathcal{T}_3$, namely

$$\ell(x) = \frac{\sqrt{2}\,\Gamma(3/2)/\sqrt{2\pi}}{\sqrt{x-1}x^{3/2}} \mathbb{I}_{(1,\infty)}(x) \, .$$

(You can verify that this is the correct normalizing constant by running `integrate`.)

The comparison of defensive sampling with the original importance sampler thus consists in adding a small sample from ℓ to the original sample from $g = f$:

```
> sam1=rt(.95*10^4,df=2)
> sam2=1+.5*rt(.05*10^4,df=2)^2
> sam=sample(c(sam1,sam2),.95*10^4)
> weit=dt(sam,df=2)/(0.95*dt(sam,df=2)+.05*(sam>0)*
+ dt(sqrt(2*abs(sam-1)),df=2)*sqrt(2)/sqrt(abs(sam-1)))
> plot(cumsum(h(sam1))/(1:length(sam1)),ty="l")
> lines(cumsum(weit*h(sam))/1:length(sam1),col="blue")
```

Note that simulations that are smaller than 1 get a weight equal to $1/.95$ in the defensive version since $\ell(x) = 0$ for $x \le 1$. As in Example 3.8 and Figure 3.7, the original sample may exhibit important jumps in the cumulated average that are warnings of infinite variance problems. The defensive sampling solution produces a much more stable evaluation of the integral. In alternative simulations, both convergence graphs may also be quite similar if no simulation is close enough to 1 to induce a large value of $1/\sqrt{x-1}$. In Figure 3.8, we selected one occurrence of discrepancy between both samples where defensive sampling brings a clear element of stabilization.   ◀

The example above clearly illustrates the impact of defensive sampling when the heavy-tailed component of the mixture is somehow related to the problem at hand. Generic choices of ℓ often lead to less efficient solutions, even when they ensure a finite variance for the Monte Carlo estimator.

Example 3.10. This example considers a probit model from a Bayesian point of view. We recall that the probit model is a particular case of a generalized linear

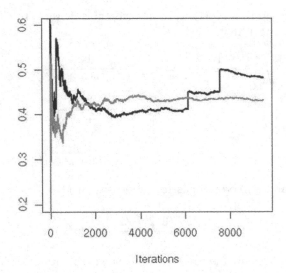

Fig. 3.8. Convergence of two estimators of the integral (3.9) of Example 3.9 based on a sample from $\mathcal{T}_2$ (dark line) and a defensive version (grey line).

model where the observables y are binary variables, taking values 0 and 1, and the covariates are vectors $x \in \mathbb{R}^p$ such that

$$\Pr(y = 1|x) = 1 - \Pr(y = 0|x) = \Phi(x^\mathsf{T}\beta), \quad \beta \in \mathbb{R}^p.$$

Data for this model can easily be simulated, but we use instead an R dataset called `Pima.tr` that is available in the library `MASS`. This dataset surveys 200 Pima Indian women in terms of presence or absence of diabetes, `Pima.tr$type` (this is the binary variable y to explain) and various physiological covariates. For illustration purposes, we only consider the body mass index covariate, `Pima.tr$bmi`, with an intercept.

A standard GLM estimation of the model is provided by

```
> glm(type~bmi,data=Pima.tr,family=binomial(link="probit"))
```

```
Coefficients:
            Estimate Std. Error z value Pr(>|z|)
(Intercept) -2.54303    0.54211  -4.691 2.72e-06 ***
bmi          0.06479    0.01615   4.011 6.05e-05 ***
---
Signif. codes: 0 '***' 0.001 '**' 0.01 '*' 0.05 '.' 0.1 ' ' 1
```

which indicates that the body mass index covariate has a significant impact on the possible presence of diabetes.

From a Bayesian perspective, we introduce a vague prior on $\beta = (\beta_1, \beta_2)$ that is a normal $\mathcal{N}(0, 100)$ distribution. The posterior distribution on β is then the product of this essentially flat prior by the likelihood, which can be defined as

```
like=function(beda){
  mia=mean(Pima.tr$bmi)
  prod(pnorm(beda[1]+(Pima.tr$bm[Pima.tr$t=="Yes"]-
    mia)*beda[2]))*
  prod(pnorm(-beda[1]-(Pima.tr$bm[Pima.tr$t=="No"]
    -mia)*beda[2]))/exp(sum(beda^2)/200)
}
```

Experimenting with the image function and this likelihood indicates that the central part of the likelihood is located near the maximum likelihood estimator (MLE) with a range of $-.6/-.3$ for the intercept β_1 and a range of $0.04/0.09$ for β_2. Using a normal proposal centered at the MLE with a diagonal covariance matrix corresponding to the estimate provided by glm is a natural choice for g, even though this does not guarantee a finite variance for all purposes. However, implementing this idea with

```
> sim=cbind(rnorm(10^3,mean=-.4,sd=.04),
+           rnorm(10^3,mean=.065,sd=.005))
> weit=apply(sim,1,post)/(dnorm(sim[,1],mean=-.4,sd=.04)*
+           dnorm(sim[,2],mean=.065,sd=.005))
```

shows that the importance weights are rather uneven, if not degenerate (you can check using boxplot(weit) for instance). A representation of 10^4 resampled points based on those weights in Figure 3.9 confirms this pattern.

In order to evaluate the (low) impact of a defensive sampling implementation, we also create an importance sample that includes simulations from the prior with probability .05 by modifying the above into

```
> sim=rbind(sim[1:(.95*10^3),],cbind(rnorm(.05*10^3,sd=10),
+          rnorm(.05*10^3,sd=10)))
> weit=apply(sim,1,post)/(.95*dnorm(sim[,1],m=-.4,sd=.081)*
+    dnorm(sim[,2],m=0.065,sd=.01)+.05*dnorm(sim[,1],sd=10)*
+    dnorm(sim[,2],sd=10))
```

The difference in efficiency is not visible, though. When use the effective sample size criterion (defined in Section 4.4), the difference on 10^3 simulations is an effective sample size of 302 for the normal sample versus an effective sample size of 283 for the defensive one. The estimates of β produced by both methods are $(-0.452, .0653)$ and $(-0.452, .0652)$, respectively. (Note the proximity with the MLE if we incorporate the mean of Pima.tr$bmi.) The reason for this strong similarity is that the additional term in the denominator due to the inclusion of the prior density is mostly zero. ◀

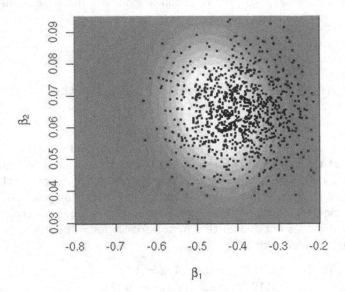

Fig. 3.9. Posterior distribution of the parameter (β_1, β_2) for the regression of diabetes on body mass index in the Pima.tr dataset with resampled values from a normal proposal superimposed.

3.4 Additional exercises

Exercise 3.9 For the same estimator $\delta(x)$ as in Exercise 3.1:

a. Build an Accept–Reject algorithm based on a Cauchy candidate to generate a sample from the posterior distribution and then deduce the estimator.
b. Design a computer experiment to compare Monte Carlo errors when using (i) the same random variables θ_i in the numerator and denominator or (ii) different random variables.

Exercise 3.10 Consider the same question as in Exercise 3.6 when

$$\omega_i = \lfloor f(X_i)/g(X_i) \rfloor + \delta_i, \text{ with } \delta_i \sim \mathcal{B}\text{in}\{1, f(X_i)/g(X_i) - \lfloor f(X_i)/g(X_i) \rfloor\}$$

and $\lfloor x \rfloor$ denoting the integer part of x. Show that there also exists an unbiased estimator based on the replacement of $f(X_i)/g(X_i)$ with $\alpha\, f(X_i)/g(X_i)$ for any $\alpha > 0$.

Exercise 3.11 Referring to Example 3.5:

a. Show that to simulate $Y \sim \mathcal{E}xp^+(a, 1)$, an exponential distribution left truncated at a, we can simulate $X \sim \mathcal{E}xp(1)$ and take $Y = a + X$.
b. Use this method to calculate the probability that a χ_3^2 random variable is greater than 25 and that a t_5 random variable is greater than 50.

c. Explore the gain in efficiency from this method. Take $a = 4.5$ in part (a) and run an experiment to determine how many random variables would be needed to calculate $P(Z > 4.5)$ to the same accuracy obtained from using 100 random variables in an importance sampler.

Exercise 3.12 Show that if

$$
\omega_i \sim \begin{cases} \mathcal{B}in(1, f(X_i)/g(X_i)) & \text{if } f(X_i)/g(X_i) < 1, \\ \mathcal{G}eo(g(X_i)/f(X_i)) & \text{otherwise}, \end{cases}
$$

the estimator $\frac{1}{n}\sum_{i=1}^{n} \omega_i h(x_i)$ is also unbiased.

Exercise 3.13 (Ó Ruanaidh and Fitzgerald, 1996) For simulating random variables from the density $f(x) \propto \exp\{-x^2\sqrt{x}\}[\sin(x)]^2$, $0 < x < \infty$, compare the following choices of instrumental densities on $\mathbb{R}$:

$$
g_1(x) = \frac{1}{2}e^{-|x|}, \quad g_2(x) = \frac{1}{2\pi}\frac{1}{1+x^2/4}, \quad g_3(x) = \frac{1}{\sqrt{2\pi}}e^{-x^2/2}.
$$

For each of the instrumental densities, estimate the number M of simulations needed to obtain three digits of accuracy in estimating $\mathbb{E}_f[X]$. Deduce from the acceptance rate an estimator of the normalizing constant of f for each of the instrumental densities.

Exercise 3.14 When a cdf $F(x)$ has a tail power of α (i.e., when $1 - F(x) \propto x^{-\alpha}$ for x large enough):

a. Show that $\mathbb{E}[X|X > K] = K\alpha/(\alpha - 1)$ for K large enough. Discuss the existence of this expectation as a function of α.
b. Derive an estimate of $\mathbb{E}[X|X > K]$ based on a sample from F.
c. Evaluate the stability of this estimate as a function of K when F is a Pareto $\mathcal{P}(2)$, $\mathcal{P}(3)$, $\mathcal{P}(4)$ distribution (see Exercise 2.13).

Exercise 3.15 (Gelfand and Dey, 1994) Consider a density function $f(x|\theta)$ and a prior distribution $\pi(\theta)$ such that the marginal $m(x) = \int_\Theta f(x|\theta)\pi(\theta)d\theta$ is finite a.e. The marginal density is of use in the comparison of models since it appears in the Bayes factor (see Robert, 2001).

a. Give the general shape of an importance sampling approximation of m.
b. Detail this approximation when the importance function is the posterior distribution and when the normalizing constant is unknown.
c. Show that, for a proper density τ,

$$
m(x)^{-1} = \int_\Theta \frac{\tau(\theta)}{f(x|\theta)\pi(\theta)}\pi(\theta|x)d\theta,
$$

and deduce that when the θ_i^*'s are generated from the posterior,

$$
\hat{m}(x) = \left\{\frac{1}{T}\sum_{t=1}^{T}\tau(\theta_i^*)\bigg/ f(x|\theta_i^*)\pi(\theta_i^*)\right\}^{-1}
$$

is another importance sampling estimator of $m(x)$.

Exercise 3.16 Given a real importance sample $X_1, \ldots, X_n$ with importance function g and target density f:

a. Show that the sum of the weights $w_i = f(X_i)/g(X_i)$ is only equal to n in expectation and deduce that the weights need to be renormalized even when both densities have known normalizing constants.
b. Assuming that the weights w_i have been renormalized to sum to one, we sample, with replacement, n points $\tilde{X}_j$ from the X_i's using those weights. Show that the $\tilde{X}_j$'s satisfy

$$\mathbb{E}\left[\frac{1}{n}\sum_{j=1}^{n} h(\tilde{X}_j)\right] = \mathbb{E}\left[\sum_{i=1}^{n} w_i h(X_i)\right].$$

c. Deduce that if the formula above is satisfied for $w_i = f(X_i)/g(X_i)$ instead, the empirical distribution associated with the $\tilde{X}_j$'s is unbiased.

Exercise 3.17 Monte Carlo marginalization is a technique for calculating a marginal density when simulating from a joint density. Let $(X_i, Y_i) \sim f_{XY}(x,y)$, independent, and the corresponding marginal distribution $f_X(x) = \int f_{XY}(x,y)dy$.

a. Let $w(x)$ be an arbitrary density. Show that

$$\lim_n \frac{1}{n}\sum_{i=1}^{n} \frac{f_{XY}(x^*, y_i)w(x_i)}{f_{XY}(x_i, y_i)} = \int\int \frac{f_{XY}(x^*, y)w(x)}{f_{XY}(x, y)} f_{XY}(x,y)dxdy = f_X(x^*),$$

which provides a Monte Carlo estimate of f_X, the marginal distribution of X, when the joint distribution is only known up to a constant.
b. Let $X|Y = y \sim \mathcal{G}(y, 1)$ and $Y \sim \mathcal{E}xp(1)$. Use the technique above to plot the marginal density of X. Compare this with the exact marginal.
c. Show that choosing $w(x) = f_X(x)$ works to produce the marginal distribution and that it is optimal in the sense of the variance of the resulting estimator.

Exercise 3.18 Given the *Gumbel distribution*, with density $f(x) = \exp\{x - \exp(x)\}$ over the real line, we are interested in comparing the variability of regular importance sampling based on a normal importance function with the variability of the corresponding self-normalized version of (3.7).

a. Show that the expectation of $\exp(X)$ is well-defined for the Gumbel distribution.
b. Create a matrix x of normal simulations with 100 columns using rnorm(100*10^4) and deduce the importance weights we.
c. Deduce the regular and the self-normalized sequences of estimators of $\mathbb{E}[\exp(X)]$ by

```
> nore=apply(we*exp(x),2,cumsum)/(1:10^4)
> reno=apply(we*exp(x),2,cumsum)/apply(we,2,cumsum)
```

and plot the ranges of both sequences of estimates using polygon.

Exercise 3.19 (Berger et al., 1998) For a $p \times p$ positive-definite symmetric matrix Σ, consider the distribution

$$\pi(\theta) \propto \exp\left(-(\theta - \mu)^t \Sigma^{-1}(\theta - \mu)/2\right)\bigg/||\theta||^{p-1}.$$

a. Show that the distribution is well-defined; that is, $\int_{\mathbb{R}^p} \pi(\theta)d\theta < \infty$.
b. Show that an importance sampling implementation based on the normal instrumental distribution $\mathcal{N}_p(\mu, \Sigma)$ is not satisfactory from both theoretical and practical points of view.
c. Examine the alternative based on a gamma distribution $\mathcal{G}(\alpha, \beta)$ on $\eta = ||\theta||^2$ and a uniform distribution on the angles.

4

Controlling and Accelerating Convergence

Bartholomew smiled. "Just because we cannot find the link here and now does not mean that it is not there. The evidence we have at the moment is just not sufficient to support any firm conclusion."

Susanna Gregory
An Unholy Alliance

Reader's guide

In Chapter 3, the Monte Carlo method was introduced (and discussed) as a simulation-based approach to the approximation of complex integrals. While the principles should by now be well-understood, there is more to be said about convergence assessment; that is, when and why to stop running simulations. We present in this chapter the specifics of variance estimation and control for Monte Carlo methods, as well as accelerating devices. We particularly focus in Sections 4.2 and 4.5 on the construction of confidence bands, stressing the limitations of normal-based evaluations in Section 4.2 and developing variance estimates for importance samplers in Section 4.3 and convergence assessment tools in Section 4.4. These are fundamental concepts, and we will see connections with similar developments in the realm of MCMC algorithms, which are discussed in Chapters 6–8. The second part of the chapter covers various accelerating devices such as Rao–Blackwellization in Section 4.6 and negative correlation in Section 4.7.

C.P. Robert, G. Casella, *Introducing Monte Carlo Methods with R*, Use R,
DOI 10.1007/978-1-4419-1576-4_4, © Springer Science+Business Media, LLC 2010

4.1 Introduction

Chapter 3 mentioned that the Central Limit Theorem applies to Monte Carlo estimates of the form

$$\overline{h}_n = \frac{1}{n} \sum_{j=1}^{n} h(X_j) \qquad X_j \sim f(x)$$

(under integrability conditions) and thus that it can be exploited for assessing the convergence to the integral of interest,

(4.1)
$$\mathfrak{I} = \int h(x) f(x) \, dx,$$

in the sense that the random variable $\sqrt{n}(\overline{h}_n - \mathfrak{I})$ is asymptotically normal. The lower panel of Figure 3.3 associated with Example 3.3 provides a direct illustration of the use of a normal confidence interval for this assessment. It represents, for each fixed number of iterations, an asymptotically valid confidence interval on the value of (4.1).

However, this approach has limitations in that the envelope built over iterations and represented in Figure 3.3 has no overall validity as a *confidence band*. Indeed, if you repeat the Monte Carlo experiment once again, the sequence $(\overline{h}_n)$ produced on the second run will most likely not stay within this envelope, and if you repeat it many times, the frequency with which it will stay within the band will not meet the nominal probability of 0.95.

The explanation for this apparent discrepancy is that the monitoring method illustrated by the lower panel of Figure 3.3 is essentially univariate in nature. That is, the confidence bounds placed on the estimate $\overline{h}_k$ at iteration k only depend on the values of $\overline{h}_k$ and the variance estimate at time k, and they thus ignore any correlation structure in the iterates. A valid confidence band needs to account for the distribution of the entire sequence, in a multivariate or even functional perspective, as discussed in Section 4.5, even though the theoretical construction of such a global band is somehow beyond our reach.

We stress that the confidence band provided by the Central Limit Theorem is asymptotically valid. What we are seeking is a more global assessment of convergence for the sequence of estimators as the number of simulations varies. This can be considered as a second-order convergence assessment, necessarily more conservative—i.e. requiring more simulations—than the original one. Although we do not male the link explicit here, the fixed-width batch mean and the bootstrap methods of Jones et al. (2006) described in Section 8.4.4 also apply to the iid settings of the current chapter.

4.2 Monitoring Variation

First, you must realize there is a straightforward and simple solution for evaluating the variability of a sequence of Monte Carlo estimates, which is to run several independent sequences in parallel. This is both easier to derive and more widely applicable than the techniques based on asymptotic approximation, if much greedier in computing time. Unfortunately, this last characteristic is a feature we will meet repeatedly in the book, namely that validation of the assessment of variation is of a higher order than convergence of the estimator itself. Namely, error assessment requires much more computing time than validation of the pointwise convergence (except in very special cases such as regeneration, covered in Robert and Casella, 2004). An approximate but cheaper version of this basic Monte Carlo estimate of the variability is to bootstrap (see Section 1.5) the current sample, as already used in Example 3.7 (see the lower left panel of Figure 3.6).

Example 4.1. If we repeat the simulations of Example 3.3, we can produce a matrix of converging estimators as in

```
> x=matrix(h(runif(200*10^4)),ncol=200)
> estint=apply(x,2,cumsum)/(1:10^4)
```

and thus obtain a Monte Carlo evaluation of the Monte Carlo variation by

```
> plot(estint[,1],ty="l",col=0,ylim=c(.8,1.2))
> y=apply(estint,1,quantile,c(.025,.975))
> polygon(c(1:10^4,10^4:1),c(y[1,],rev(y[2,])),col="wheat")
```

At any iteration, the band represented in Figure 4.1 contains 95% of the estimation sequences. Obviously, if we pick any of the convergence sequences thus produced, the CLT confidence band will fail to correspond to this overall band since, as in Figure 3.3 *(lower)*, it will reproduce the variations of the original sequence. If we now consider the bootstrapped version of the overall confidence band, we start by producing bootstrapped replicas of the original sequence x[,1] using

```
> boot=matrix(sample(x[,1],200*10^4,rep=T),nrow=10^4,ncol=200)
```

and then reproduce the confidence band construction by

```
> bootit=apply(boot,2,cumsum)/(1:10^4)
> bootup=apply(bootit,1,quantile,.975)
> bootdo=apply(bootit,1,quantile,.025)
```

As shown in Figure 4.1, the band thus produced has a behavior that is quite similar to that of the band resulting from iid replications of the Monte Carlo sequence, except for a drift in its location. The gain in using the bootstrap version is obviously that only a single sequence needs to be produced.   ◄

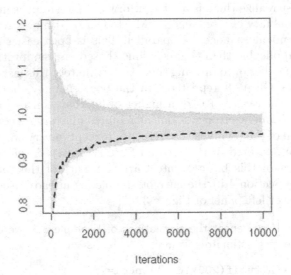

Fig. 4.1. Convergence band of 200 parallel sequences of Monte Carlo estimators of the integral of $h(x) = [\cos(50x) + \sin(20x)]^2$ *(lower)* and bootstrap version based on a single sequence *(upper)*, with its lower bound represented by a dotted line.

In Example 4.1, the appeal of using a bootstrapped confidence band is somehow limited because the computing cost of producing a bootstrap sequence is approximately the same as the computing cost of producing a new sequence. In more complex settings, however, producing a new sequence may prove much more costly than resampling from the original sequence.

This simple example thus warns us against the blind use of a normal approximation when repeatedly invoked over iterations with dependent estimators simply because the normal confidence approximation only has a pointwise validation. Using a band of estimators in parallel is obviously more costly, but it provides the correct assessment on the variation of these estimators.

4.3 Asymptotic variance of importance sampling estimators

The following example illustrates a basic difficulty when assessing convergence for importance sampling.

Example 4.2. For a normal observation equal to $x = 2.5$ with a Cauchy prior distribution on its mean,

$$X \sim \mathcal{N}(\theta, 1), \quad \theta \sim \mathcal{C}(0, 1),$$

the posterior expectation of the mean θ is given by

$$\delta^{\pi}(x) = \int_{-\infty}^{\infty} \frac{\theta}{1+\theta^2} e^{-(x-\theta)^2/2} d\theta \bigg/ \int_{-\infty}^{\infty} \frac{1}{1+\theta^2} e^{-(x-\theta)^2/2} d\theta.$$

Therefore, an approximation of $\delta^{\pi}(x)$ can be based on the simulation of iid variables $\theta_1, \cdots, \theta_n \sim \mathcal{N}(x, 1)$ as

(4.2)
$$\hat{\delta}_n^{\pi}(x) = \sum_{i=1}^n \frac{\theta_i}{1+\theta_i^2} \bigg/ \sum_{i=1}^n \frac{1}{1+\theta_i^2}$$

since both the numerator and the denominator are convergent (in n). Note that the estimator can also be interpreted as a self-normalized importance sampling approximation with the importance function being the normal density and the importance ratio being equal to $1/(1+\theta^2)$. ◀

A difficulty associated with this example (and any other self-normalized importance sampler) is that the estimator $\hat{\delta}_n^{\pi}(x)$ is in fact a ratio of estimators and that the variance of a ratio is *not* the ratio of the variances. This is a common occurrence in Bayesian computations (see Exercises 4.1 and 4.2).

⚡ As mentioned previously, normalizing constants are superfluous in Bayesian inference, *except* in the case when several models are considered simultaneously in order to be compared. In those occurrences, the fundamental quantity is the Bayes factor

$$\rho = \frac{m_1(x)}{m_2(x)} = \frac{\int_{\Theta_1} \pi_1(\theta_1) f_1(x|\theta_1) \, d\theta_1}{\int_{\Theta_2} \pi_2(\theta_2) f_2(x|\theta_2) \, d\theta_2}$$

whose position against 1 drives the comparison (see, e.g., Robert, 2001, Chapters 5 and 6). In this case, the posterior distributions for both models under comparison are typically available up to a normalizing constant,

$$\pi_1(\theta|x) = \tilde{\pi}_1(\theta)/c_1 \text{ and } \pi_2(\theta|x) = \tilde{\pi}_2(\theta)/c_2,$$

where only $\tilde{\pi}_1$ and $\tilde{\pi}_2$ are known and where c_1 and c_2 correspond to the marginal likelihoods, $m_1(x)$ and $m_2(x)$ (the dependence on x is removed for simplification purposes). The Bayes factor is therefore identical to the ratio of those missing constants, $\varrho = c_1/c_2$. Special computational techniques have been devised for the approximation of Bayes factors, as in Chen et al. (2000) (see also Exercise 4.2).

Exercise 4.1 We assume here that both posteriors are absolutely continuous with respect to one another (i.e., that the parameters for both models live in the same space).

a. Show that the Bayes factor ϱ can be approximated by

$$\hat{\varrho} = \frac{1}{n}\sum_{i=1}^{n} \frac{\tilde{\pi}_1(\theta_i)}{\tilde{\pi}_2(\theta_i)}, \qquad \theta_1,\ldots,\theta_n \sim \pi_2.$$

b. Show that the identity

$$\frac{\int \tilde{\pi}_1(\theta)\alpha(\theta)\pi_2(\theta|x)d\theta}{\int \tilde{\pi}_2(\theta)\alpha(\theta)\pi_1(\theta|x)d\theta} = \frac{c_1}{c_2} = \varrho$$

holds for every function α such that both integrals are finite.

The transform $\xi = \log(\varrho)$ is often seen as more relevant for the model comparison (because of the χ^2 scale of the log-likelihood ratio), and this log-odds ratio can be approximated in its own right.

Exercise 4.2 Under the same assumption as in Exercise 4.1, when the priors π_1 and π_2 belong to the same parameterized family (that is, when $\pi_i(\theta) = \pi(\theta|\lambda_i)$), the corresponding normalizing constants are denoted by $c(\lambda_i)$. The parameter λ is a real number.

a. When $\pi(\lambda)$ is an arbitrary distribution on λ with support (λ_1, λ_2), verify the identity

$$-\log\left(\frac{c(\lambda_1)}{c(\lambda_2)}\right) = \mathbb{E}\left[\frac{U(\theta,\lambda)}{\pi(\lambda)}\right],$$

where

$$U(\theta,\lambda) = \frac{d}{d\lambda}\log(\tilde{\pi}(\theta|\lambda))$$

and the expectation is computed under $\pi(\theta|\lambda)\pi(\lambda)$.

b. Deduce that ξ can be estimated with the *path sampling estimator*

$$\hat{\xi} = \frac{1}{n}\sum_{i=1}^{n} U(\theta_i, \lambda_i) \Big/ \pi(\lambda_i)$$

when the (θ_i, λ_i)'s are simulated from the joint distribution $\pi(\lambda)\pi(\theta|\lambda)$.

c. Show that the minimum variance estimator of ξ over all $\pi(\lambda)$'s is based on

$$\pi^\star(\lambda) \propto \sqrt{\mathbb{E}[U^2(\theta,\lambda)|\lambda]}.$$

Consider thus a general ratio estimator

$$\delta_h^n = \sum_{i=1}^{n} \omega_i \, h(x_i) \Big/ \sum_{i=1}^{n} \omega_i,$$

where the x_i's are realizations of random variables $X_i \sim g(y)$ with g as a candidate distribution for target f. In addition, the ω_i's are realizations of random variables W_i such that $\mathbb{E}[W_i | X_i = x] = \kappa f(x)/g(x)$, κ being an arbitrary constant (that corresponds to the lack of normalizing constants in f and g). We denote

$$S_h^n = \sum_{i=1}^{n} W_i h(X_i), \quad S_1^n = \sum_{i=1}^{n} W_i.$$

(Note that we do not assume independence between the X_i's as in regular importance sampling.) Then, as shown in Liu (1996) and Robert and Casella (2004, Chapter 4), the asymptotic variance of δ_h^n is

$$\text{var}(\delta_h^n) = \frac{1}{n^2 \kappa^2} \left\{ \text{var}(S_h^n) - 2\mathbb{E}_f[h] \, \text{cov}(S_h^n, S_1^n) + \mathbb{E}_f[h]^2 \, \text{var}(S_1^n) \right\}.$$

Following Liu (1996), we can then deduce that, when considering iid X_i's, the self-normalized importance sampling estimator, and the right degree of approximation, a rough approximation to its variance is

$$\text{var}(\delta_h^n) \approx \frac{1}{n} \text{var}_f(h(X)) \left\{ 1 + \text{var}_g(W) \right\}.$$

⚡ The approximation above is only valid for the *normalized* version of the weights ω_i, for otherwise it would depend on the missing constant κ. Its Monte Carlo estimate is thus

$$(4.3) \qquad \frac{\sum_{i=1}^{n} \omega_i \left\{ h(x_i) - \delta_h^n \right\}^2}{n \sum_{i=1}^{n} \omega_i} \left\{ 1 + n^2 \, \widehat{\text{var}}(W) \Big/ \left(\sum_{i=1}^{n} \omega_i \right)^2 \right\},$$

where $\widehat{\text{var}}$ denotes the standard variance estimate (as when using var). This expression and its simplification will be further discussed in Section 4.4 when building a convergence assessment based solely on the weights.

This expression of the variance is a rather crude approximation, as can be seen through the fact that this quantity is always larger than $\text{var}_f(h(X))$, which provides the variance for an iid sample with the same sample size. Since there exist choices of g for which the variance $\text{var}(\delta_h^n)$ is exactly equal to 0 (see Robert and Casella, 2004), this ordering cannot always hold.

Example 4.3. (Continuation of Example 4.2) If we generate a normal $\mathcal{N}(x, 1)$ sample for the importance sampling approximation (4.2), the variance approximation above can be used to assess the variability of these estimates, but, again, the asymptotic nature of the approximation must be taken into account. If we take as reference the range of 500 parallel sequences of estimators of $\delta^{\pi}(x)$,

```
> norma=matrix(rnorm(500*10^4),ncol=500)+2.5
> weit=1/(1+norma^2)
> esti=apply(norma*weit,2,cumsum)/apply(weit,2,cumsum)
> plot(esti[,1],type="l",col="white",ylim=c(1.7,1.9))
> band=apply(esti,1,quantile,c(.025,.975))
> polygon(c(1:10^4,10^4:1),c(band[1,],rev(band[2,])))
```

the juxtaposition of the band produced on a single sequence shows an underestimation of the variation if the usual variance estimate is used

```
> vare=cumsum(weit[,1]*norma[,1]^2)/
+       cumsum(weit[,1])-esti[,1]^2
> lines(esti[,1]+2*sqrt(vare/(1:10^4)))
> lines(esti[,1]-2*sqrt(vare/(1:10^4)))
```

and an equivalent if drifted range for the weight correction

```
> varw=cumsum(weit[,1]^2)*(1:10^4)/cumsum(weit[,1])^2
> lines(esti[,1]+2*sqrt(varw*vare/(1:10^4)),col="sienna")
> lines(esti[,1]-2*sqrt(varw*vare/(1:10^4)),col="sienna")
```

since the respective ranges are 0.0559 for the former (Monte Carlo) and 0.0539 for the latter (corrected variance). Figure 4.2 compares the approximated variance bands for $x = 2.5$ with the actual variation of the estimates, evaluated over the 500 parallel sequences.

Figure 4.3 reproduces this evaluation for θ_i's simulated from the prior $\mathcal{C}(0, 1)$ distribution and for the importance sampling estimate[1]

$$\tilde{\delta}_h^n = \sum_{i=1}^{n} \theta_i \exp\left\{-(x - \theta_i)^2/2\right\} \Big/ \sum_{i=1}^{n} \exp\left\{-(x - \theta_i)^2/2\right\}.$$

In this case, since the corresponding functions h are bounded for both choices, the variabilities of the estimates are quite similar, with a slight advantage to the normal sampling. The range of the Monte Carlo evaluation is indeed 0.0594, while the range of the asymptotic band is 0.0800. The fact that the latter is based on a *single* sequence must be taken into account. Another sequence would produce a different range, as you can (and should) easily check. Similarly, the boundary curves appearing in both Figures 4.2 and 4.3 are produced by a single sequence and thus should not be overinterpreted.   ◄

---

[1] The inversion of the roles of the $\mathcal{N}(x, 1)$ and $\mathcal{C}(0, 1)$ distributions illustrates once more both the ambiguity of the integral representation and the opportunities offered by importance sampling.

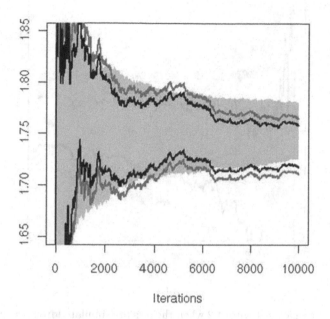

Iterations

Fig. 4.2. Convergence of a sequence of Monte Carlo estimators for the posterior mean in the Cauchy-normal problem when $x = 2.5$ and the simulations are from a normal $\mathcal{N}(x, 1)$ importance function. The shaded zone represents the 95% confidence range on the entire set of 500 parallel sequences of Monte Carlo estimators at each iteration, the inner boundary corresponds to the normal band for the standard variance estimate, and the outer (lighter) boundary corresponds to the normal band for the corrected variance estimate.

In the example above, running the approximation with a Cauchy sample leads to an interesting if minor implementation problem. Since the Cauchy distribution has very heavy tails, some of the 500*10^4 simulations take very large values, which then leads to weights equal to zero in the R output, even though the normal density is formally strictly positive:

```
> cocha=matrix(rcauchy(500*10^4),ncol=500)
> range(cocha)
[1] -18228407   3461090
> wach=dnorm(cocha,mean=2.5)
> range(wach)
[1] 0.0000000 0.3989423
```

Since using values with zero weights as first simulations is impossible when resorting to cumsum to monitor the convergence of the self-normalized impor-

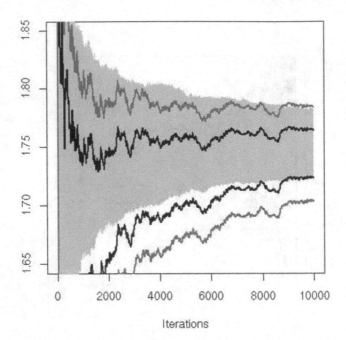

Iterations

Fig. 4.3. Same plot as Figure 4.2 when the θ_i's are simulated from the prior $\mathcal{C}(0,1)$ distribution.

tance sampling estimate, given that it produces NA due to divisions by zero, a fix to the problem is to impose a minimum value on the weights in the denominator in the first simulation, namely

```
> wachd=wach
> wachd[apply(wachd,2,cumsum)<10^(-10)]=10^(-10)
```

which is the solution used to produce Figure 4.3. Note that, except for this difficulty, the behavior of the weights is quite similar in both approaches, as can be checked via boxplot.

4.4 Effective sample size and perplexity

The representation (4.3) proposed for the approximate variance of the self-normalized importance sampling estimator leads to an essential tool for assessing the performance of importance samplers. When expanding the empirical variance $\widehat{\text{var}}(W)$,

$$\widehat{\mathrm{var}}(W) = \frac{1}{n} \sum_{i=1}^{n} \omega_i^2 - \frac{1}{n^2} \left(\sum_{i=1}^{n} \omega_i \right)^2,$$

the coefficient $1 + \widehat{\mathrm{var}}_g(W)$ is equal to

$$n^2 \sum_{i=1}^{n} \omega_i^2 \Big/ \left(\sum_{i=1}^{n} \omega_i \right)^2.$$

If we now denote the normalized weights by

$$\underline{\omega}_i = \omega_i \Big/ \sum_{j=1}^{n} \omega_j,$$

we then define the *effective sample size* by

$$\mathrm{ESS}_n = 1 \Big/ \sum_{i=1}^{n} \underline{\omega}_i^2.$$

Beside being useful in assessing the loss in variance due to the importance weights, this factor gives a direct evaluation of the worth of the importance sampler, as it is equivalent to a sample size. For a uniformly weighted sample, ESS_n is equal to n, while, for a completely degenerated sample where all importance weights but one are zero, ESS_n is equal to 1. The effective sample size thus evaluates the size of the iid sample equivalent to the weighted sample and allows a direct comparison of samplers.

Exercise 4.3 Show that ESS_n always takes values between 1 and n.

A second (and almost equivalent) assessment is provided by the so-called *perplexity* (Cappé et al., 2008), $\exp(\mathfrak{H}_n)/n$, where

$$\mathfrak{H}_n = -\sum_{i=1}^{n} \underline{\omega}_i \log(\underline{\omega}_i)$$

is the Shannon entropy of the normalized importance sampling weights. (This tool is primarily used in information theory and speech recognition, see Jelinek, 1999, but it nonetheless brings an additional feature to the effective sample size that can be exploited in all settings.) The perplexity provides an estimate of $\exp[\mathfrak{E}(f,g)]$, where

$$\mathfrak{E}(f,g) = \int \log \left\{ \frac{f(x)}{g(x)} \right\} f(x) \, \mathrm{d}x$$

is the Kullback–Leibler divergence between the target f and the importance function g. Therefore, the closer the perplexity is to 1, the more appropriate the importance function.

Exercise 4.4 Show that $\exp(\mathfrak{H}_n)$ always takes values between 1 and n.

Example 4.4. (Continuation of Example 4.3) If we compare the effective sample sizes for the normal and the Cauchy simulations, the difference in efficiency (and perplexity) is much clearer than in the comparison between Figures 4.2 and 4.3. The computation of those quantities is straightforward. For the effective sample sizes,

```
> ess=apply(weit,2,cumsum)^2/apply(weit^2,2,cumsum)
> essbo=apply(ess,1,quantile,c(.025,.975))
> ech=apply(wachd,2,cumsum)^2/apply(wachd^2,2,cumsum)
> echbo=apply(ech,1,quantile,c(.025,.975))
```

and for the perplexities,

```
> sumweit=apply(weit,2,cumsum)
> plex=(apply(weit*log(weit),2,cumsum)/sumweit)-log(sumweit)
> chumweit=apply(wachd,2,cumsum)
> plech=(apply(wachd*log(wachd),2,cumsum)/chumweit)-
+       log(chumweit)
> plob=apply(exp(plex),1,quantile,c(.025,.975))
> ploch=apply(exp(plech),1,quantile,c(.025,.975))
```

As shown by both panels of Figure 4.4, the normal-based simulation is clearly twice as efficient as the Cauchy-based scenario. Note the strong stability of those indicators across sequences.   ◀

4.5 Simultaneous monitoring

As mentioned in the introduction to this chapter, one valid method for attaching variances to a mean plot, and thus for building a valid Central Limit Theorem, is to derive the bounds using a multivariate approach. A first construction, based on a normal approximation, is found in Robert and Casella (2004, Section 4.1.2).

Given an iid sequence $X_1, X_2, \ldots, X_n$ with mean $\mu = \mathbb{E}_f(X_1)$, we want to produce an error bound on the sequence $(\bar{X}_m)_{1 \leq m \leq n}$, where $\bar{X}_m = (1/m)\sum_{i=1}^m X_i$. If, further, $X_i \sim \mathcal{N}(\mu, \sigma^2)$, then

$$\bar{\mathbf{X}} = (\bar{X}_1, \bar{X}_2, \ldots, \bar{X}_n) \sim \mathcal{N}_n(\mathbf{1}\mu, \boldsymbol{\Sigma}),$$

with

$$\boldsymbol{\Sigma} = \sigma^2 \begin{pmatrix} 1 & \frac{1}{2} & \frac{1}{3} & \frac{1}{4} & \frac{1}{5} & \cdots & \frac{1}{n} \\ \frac{1}{2} & \frac{1}{2} & \frac{1}{3} & \frac{1}{4} & \frac{1}{5} & \cdots & \frac{1}{n} \\ \frac{1}{3} & \frac{1}{3} & \frac{1}{3} & \frac{1}{4} & \frac{1}{5} & \cdots & \frac{1}{n} \\ \vdots & \vdots & \vdots & \vdots & \vdots & \cdots & \vdots \\ \frac{1}{n} & \frac{1}{n} & \frac{1}{n} & \frac{1}{n} & \frac{1}{n} & \cdots & \frac{1}{n} \end{pmatrix}.$$

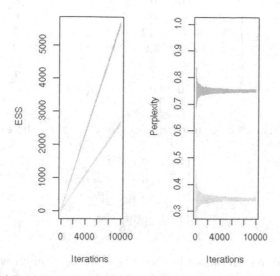

Fig. 4.4. *(left)* Evolution of the effective sample sizes (ESS) for the normal *(grey)* and the Cauchy *(light grey)* importance weights; *(right)* evolution of the perplexities with the same color codes. (In both graphs, the normal importance sampler is above the Cauchy importance sampler.)

Exercise 4.5 Establish that

$$\operatorname{cov}(\bar{\mathbf{X}}_k, \bar{\mathbf{X}}_{k'}) = o^2/\max\{k, k'\}.$$

Since

$$(4.4) \qquad (\bar{\mathbf{X}} - \mathbf{1}\mu)'\Sigma^{-1}(\bar{\mathbf{X}} - \mathbf{1}\mu) \sim \begin{cases} \chi_n^2 & \text{if } \sigma^2 \text{ is known,} \\ nF_{n,\nu} & \text{if } \sigma^2 \text{ is unknown,} \end{cases}$$

we have a simultaneous variation region for the entire vector $\bar{\mathbf{X}}$ of the form

$$\left\{ \bar{\mathbf{x}} : (\bar{\mathbf{x}} - \mathbf{1}\mu)'\Sigma^{-1}(\bar{\mathbf{x}} - \mathbf{1}\mu) \leq d_n \right\},$$

where d_n is the appropriate χ_n^2 or $F_{n,\nu}$ quantile (and Σ^{-1} is based on an estimate $\hat{\sigma}^2 \sim \chi_\nu^2$ of σ^2, independent of $\bar{\mathbf{X}}$ in the latter case).

Furthermore, the inverse of Σ happens to be computable in closed form (Exercise 4.6), since it is given by

(4.5) $$\Sigma^{-1} = \frac{1}{\sigma^2} \begin{pmatrix} 2 & -2 & 0 & 0 & 0 & \cdots & & 0 \\ -2 & 8 & -6 & 0 & 0 & \cdots & & 0 \\ 0 & -6 & 18 & -12 & 0 & \cdots & & 0 \\ 0 & 0 & -12 & 32 & -20 & \cdots & & 0 \\ \vdots & \vdots & \vdots & \vdots & \vdots & \cdots & & -n(n-1) \\ 0 & 0 & 0 & 0 & \cdots & -n(n-1) & & n^2 \end{pmatrix}.$$

Exercise 4.6 Establish a recursion relation for calculating the elements a_{ij} of Σ^{-1}:

$$a_{ij} = \begin{cases} 2i^2 & \text{if } i = j < n, \\ n^2 & \text{if } i = j = n, \\ -ij & \text{if } |i - j| = 1, \\ 0 & \text{otherwise.} \end{cases}$$

Obviously, this region cannot be used per se since μ is unknown. A possible substitute is the ellipsoid centered at the final estimate $\bar{x}_n$,

$$\left\{ \bar{y} : (\bar{y} - \mathbf{1}\bar{y}_n)' \Sigma^{-1} (\bar{y} - \mathbf{1}\bar{y}_n) \leq d_n \right\},$$

but this induces a bias that should be accounted for in the bound d_n. Furthermore, the computation and representation of the confidence band induced by this condition are far from immediate. In fact, the most obvious implementations involve simulations of the same order of magnitude as the rudimentary replications proposed in Section 4.2!

Exercise 4.7 Show that the bias due to the replacement of μ by $\bar{x}_n$ is of the order of a χ_n^2 term, which can thus be corrected directly in d_n.

This procedure is, however, easy to modify into a straightforward solution with a sound theoretical[2] foundation that can be found in Kendall et al. (2007). Instead of relying on the normal approximation above. those authors relate the sequence of the X_i's to the step function δ_n defined by

$$\{\delta_n(t)\}_{t \in [0,1]} = \left(\frac{1}{\lfloor nt \rfloor} \sum_{i=1}^{\lfloor nt \rfloor} X_i \right)_{t \in [0,1]}$$

where $\lfloor x \rfloor$ denotes the integer part of x, that is, the largest integer not exceeding x. By convention, $\delta_n(t) = 0$ for $t < 1/n$. There exists a functional

---

[2] The remainder of this section requires the use of the Brownian motion. While the final outcome does not require any advanced knowledge about the Brownian motion, for those readers unfamiliar with this notion, the section can possibly be skipped since it has no bearing on the remainder of the book.

extension of the CLT called *Donsker's Theorem*, which states that both the random functional

$$\left\{ \frac{1}{\sqrt{n}} \sum_{i=1}^{\lfloor nt \rfloor} \left(\frac{X_i - \mu}{\sigma} \right) \right\}_{t \in [0,1]}$$

and its approximation

$$\left(\frac{\hat{\sigma}_n^{-1}}{\sqrt{n}} \sum_{i=1}^{\lfloor nt \rfloor} \{X_i - \mu\} \right)_{t \in [0,1]}$$

converge in distribution (when n goes to infinity) to a standard Brownian motion on $[0,1]$, denoted $\{W(t)\}_{t \in [0,1]}$ (Feller, 1971). Exploiting a confidence band u^* on the Brownian motion $\{W(t)\}_{t \in [0,1]}$ derived in the same paper, Kendall et al. (2007) then showed that the functional band

$$\left\{ \omega \, ; \, \omega(t) \in \left[\mu - u^*(t)\sqrt{n\hat{\sigma}_n^2}/\lfloor nt \rfloor, \;\; \mu + u^*(t)\sqrt{n\hat{\sigma}_n^2}/\lfloor nt \rfloor \right], \, t \in [0,1] \right\}$$

contains $(\delta_n(t))_{t \in [0,1]}$ with asymptotic probability $1 - \alpha$. While the exact form of u^* is slightly too complex to describe here, a conservative approximation at the confidence level 0.95 is provided by

$$u^*(t) = a + b\sqrt{t} \text{ with } a = 0.3 \text{ and } b = 2.35,$$

as illustrated by Figure 4.5.

Exercise 4.8 Based on the three fundamental properties of a Brownian motion $\{W(t)\}_{t \in [0,1]}$,

  i. $W(0) = 0$,
 ii. $W(t_1) - W(t_2) \sim \mathcal{N}(0, |t_1 - t_2|)$ for $0 \leq t_1, t_2 \leq 1$,
iii. $W(t_1) - W(t_2)$ is independent from $W(t_2)$ for $t_1 > t_2$,

construct an algorithm that simulates the Brownian motion at a given discretization level δ, $\{W(k\delta)\}_{k=0,...,n}$, and verify by a Monte Carlo experiment that the bound above contains the Brownian motion with probability 95%.

While this construction is appealing in controlled situations when μ is known or estimated from an independent source (see Section 4.7.3), the main item of interest from a Monte Carlo point of view is rather a confidence region that should contain a (random) Monte Carlo sequence with a given (asymptotic) probability, given the current sequence $X_1, \ldots, X_n$. Formally, given an iid sample $Y_1, \ldots, Y_n$ from the same target f that is independent of the observed $X_1, \ldots, X_n$, associated with the estimators $\delta_N(1)$ (of μ) and $\hat{\sigma}_n$ (of σ), we are interested in the variations of the random function

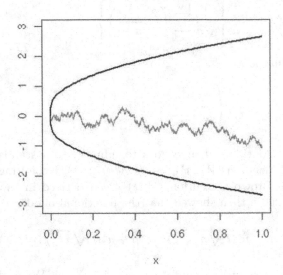

Fig. 4.5. 95% confidence bound on the Brownian motion on $[0,1]$ and representation of one realization of a Brownian motion.

$$\left(\frac{\hat{\sigma}_n^{-1}}{\sqrt{n}} \sum_{i=1}^{\lfloor nt \rfloor} \{Y_i - \delta_n(1)\}\right)_{t\in[0,1]},$$

which can be decomposed as

$$\frac{\hat{\sigma}_n^{-1}}{\sqrt{n}} \sum_{i=1}^{\lfloor nt \rfloor} \{Y_i - \delta_n(1)\} = \frac{\hat{\sigma}_n^{-1}}{\sqrt{n}} \sum_{i=1}^{\lfloor nt \rfloor} \{Y_i - \mu\} - \frac{\lfloor nt \rfloor \hat{\sigma}_n^{-1}}{\sqrt{n}} \{\delta_n(1) - \mu\}.$$

Therefore, based on Donsker's version of the CLT (Kallenberg, 2002, p. 275), the scaled cumulative estimates converge in distribution to a continuous random process that is the sum of a standard Brownian motion $\{W(t)\}_{t\in[0,1]}$ and a random linear function $(tU)_{t\in[0,1]}$, U being a standard normal random variable independent of $\{W(t)\}_{t\in[0,1]}$. For this alternative stochastic process, the confidence band is determined by a function that is approximately $u^\star(t) = a + b\sqrt{t}$, with $a = 0.1$ and $b = 3.15$ (Kendall et al., 2007). This implies that the natural confidence band associated with a sequence $X_1, \ldots, X_n$ is given by

$$\delta_n(1) \pm \frac{\hat{\sigma}_n \sqrt{n}}{\lfloor nt \rfloor} u^\star(t), \quad t \in [0,1].$$

As is obvious from the previous formula, the confidence band derived from the Brownian approximation solely depends on the estimates of the expectation $\delta_n(1)$ and its variance $\hat{\sigma}_n^2$ at the end of the simulation run. Thus, while the formal validity of the Brownian approximation is certain, when compared with the original CLT approximation in Section 4.2, it should not be overinterpreted! In particular, computing the confidence bands associated with two different simulation runs will produce different and possibly conflicting ranges. Note also that the confidence band needs to be recomputed when more simulations are added as for instance when you realize the range is too large for your estimation purposes (which is the primary reason for using those bands!). Given that the augmented confidence band corresponds to longer sentences, it should be inflated when compared with the current version.

Example 4.5. (Continuation of Example 4.4) If we want to compare the confidence bands produced by the two importance sampling approximations based on the normal and the Cauchy samples, we simply need to produce a normal and a Cauchy sequence, respectively. Given that precision or variation is what matters, we can evaluate both the numerator and denominator separately. If we consider for instance

$$\mathfrak{I}_1 = \int \frac{1}{\pi} \frac{\theta}{1 + \theta^2} \frac{1}{\sqrt{2\pi}} \exp\left\{ -\frac{1}{2}(\theta - x)^2 \right\} \, d\theta,$$

the band for the normal simulation is obtained by

```
> Nsim=10^4
> norma=rnorm(Nsim)+2.5
> hnorm=norma*dcauchy(norma)
> munorm=mean(hnorm)
> sdnorm=sd(hnorm)
> f=function(x) (cumsum(hnorm))[round(Nsim*x)]/round(x*Nsim)
> curve(munorm+(.1+3.15*sqrt(x))*sdnorm*10^2/round(x*Nsim),
+          lwd=2,from=0,to=1)
> curve(munorm-((.1+3.15*sqrt(x))*sdnorm*10^2/round(x*Nsim)),
+          lwd=2,from=0,to=1,add=T)
> curve(f,lwd=2,from=0.001,to=1,col="steelblue",add=T)
```

while the band for the Cauchy simulation is produced the same way, replacing the first two lines by

```
> norma=rcauchy(Nsim)
> hnorm=norma*dnorm(norma-2.5)
```

Since the variance estimate is much larger for the Cauchy sample, the corresponding confidence is much larger too. Figure 4.6 represents the confidence bands for both $\mathfrak{I}_1$ and

$$\mathfrak{I}_2 = \int \frac{1}{\pi} \frac{1}{1 + \theta^2} \frac{1}{\sqrt{2\pi}} \exp\left\{ -\frac{1}{2}(\theta - x)^2 \right\} \, d\theta,$$

that are associated with a normal and a Cauchy sequence, respectively. In both cases, the confidence allocated to the normal run is higher.   ◄

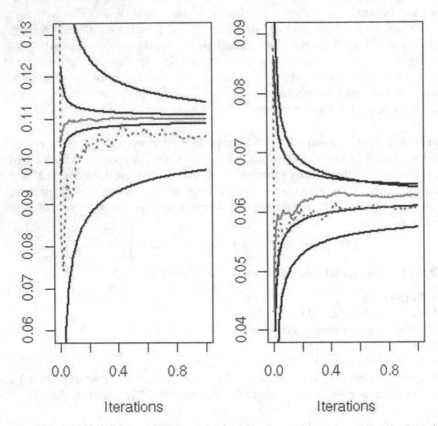

Fig. 4.6. 95% confidence bounds on the importance sampling estimates of $\mathfrak{I}_1$ *(left)* and $\mathfrak{I}_2$ *(right)* for a single normal *(full)* and a single Cauchy *(dotted)* sequence of 10^4 simulations superimposed. The Cauchy confidence bands are larger in both cases.

The construction of the confidence band above obviously applies to any transform $h(X)$ that is square-integrable. It also operates in importance sampling settings (i.e., when the Y_i's are generated from an importance density q and $h(Y_i)$ is replaced with $H(Y_i) = h(Y_i)f_\pi(Y_i)/q(Y_i)$) as well as in Markov chain Monte Carlo (MCMC) setups.

4.6 Rao–Blackwellization and deconditioning

A famous theorem of mathematical statistics, the Rao–Blackwell Theorem, states that replacing an estimator with a conditional expectation improves upon its variance,

$$\text{var}(\mathbb{E}[\delta(X)|Y]) \leq \text{var}(\delta(X)),$$

if Y is a sufficient statistic (Lehmann and Casella, 1998). This theorem also has bearings on computational methodology since it gives a generic approach to reducing the variance of a Monte Carlo estimator, which is to use conditioning. This technique is sometimes called *Rao–Blackwellization* (Gelfand and Smith, 1990, Liu et al., 1994, Casella and Robert, 1996), although the conditioning is not always in terms of sufficient statistics in Monte Carlo settings. It basically states that using conditional expectations—that can be computed—in Monte Carlo expressions brings an improvement in the variability of those Monte Carlo estimators while it does not perturb their inherent unbiasedness.

Rao–Blackwellization means taking advantage of the fact that, if $\delta(X)$ is an estimator of $\mathfrak{I} = \mathbb{E}_f[h(X)]$ and if X can be simulated from the joint distribution $f^*(x, y)$ satisfying

$$\int f^*(x, y)\, dy = f(x),$$

then the new estimator $\delta^*(Y) = \mathbb{E}_f[\delta(X)|Y]$ dominates δ in terms of variance, while the bias is the same. Obviously, this result is only useful in settings where $\delta^*(Y)$ can be explicitly computed.

Example 4.6. Consider computing the expectation of $h(x) = \exp(-x^2)$ when $X \sim \mathcal{T}(\nu, \mu, \sigma^2)$. The standard Student's t distribution has been constructed by William Gosset (see, e.g., Stigler, 1986) as the distribution of the ratio of a normal variable by a normalized χ (not χ^2!) variable; i.e., if $X \sim \mathcal{T}(\nu, \mu, \sigma^2)$, then

$$X = \mu + \sigma\frac{\epsilon}{\sqrt{\xi/\nu}}, \quad \text{with } \epsilon \sim \mathcal{N}(0, 1), \xi \sim \chi_\nu^2.$$

Therefore, even though the t distribution can be simulated directly by the function rt(), it allows for the marginal representation above in terms of the joint distribution on (x, ξ) or, equivalently (for $y = \xi/\nu$),

$$f^*(x, y) = \frac{\sqrt{y}}{\sqrt{2\pi\sigma^2}} \exp\left\{-(x - \mu)^2 y/2\sigma^2\right\} \frac{(\nu/2)^{\nu/2}}{\Gamma(\nu/2)} y^{\nu/2-1} \exp\left(-y\nu/2\right),$$

sometimes called Dickey's decomposition (1968). The implementation of the decomposition is straightforward:

```
> y=sqrt(rchisq(Nsim,df=nu)/nu)
> x=rnorm(Nsim,mu,sigma/y)
```

produces a sample of (X_i, Y_i)'s $(i = 1, \ldots, n)$. (Remember that, in R, the norm functions are parameterized by the standard deviation and not the variance!) Therefore, the empirical average

$$\delta_n = \frac{1}{n} \sum_{j=1}^{n} \exp(-X_j^2)$$

can be improved upon when the X_j's are parts of the sample $((X_1, Y_1), \ldots, (X_n, Y_n))$, since the Rao–Blackwellized version

$$\delta_n^\star = \frac{1}{n} \sum_{j=1}^{n} \mathbb{E}[\exp(-X^2)|Y_j]$$

$$= \frac{1}{n} \sum_{j=1}^{n} \frac{1}{\sqrt{2\sigma^2/Y_j + 1}} \exp\left\{-\frac{\mu^2}{1 + 2\sigma^2/Y_j}\right\}$$

can be directly computed. It is therefore a matter of comparison of

```
> d1=cumsum(exp(-x^2))/(1:Nsim)
> d2=cumsum(exp(-mu^2/(1+2*(sigma/y)^2)))/
+ sqrt(1+2*(sigma/y)^2))/(1:Nsim)
```

Figure 4.7 provides an illustration of the difference of the convergence of δ_n and $\delta_n^\star$ to $\mathbb{E}_g[\exp(-X^2)]$ for $(\nu, \mu, \sigma) = (5, 3, 0.5)$. For δ_n to have the same precision as $\delta_n^\star$ requires ten times as many simulations since the estimated variances are 0.00279 and 0.00022, respectively.   ◄

Exercise 4.9 Show that

$$\int_{-\infty}^{\infty} \exp\{-x^2\} \exp\{-(x - \mu)^2 y/2\sigma^2\} \, dx = \exp\left\{-\frac{\mu^2}{1 + 2\sigma^2/y}\right\}$$

by completing the square in the exponent to evaluate the integral.

Unfortunately, this conditioning principle seems to enjoy limited applicability since it involves a particular type of simulation (joint variables) and also requires integrand functions h that are sufficiently regular for the conditional expectations to be explicit. There are, however, specific situations where Rao–Blackwellization is always possible. One illustration is provided in Robert and Casella (2004, Section 4.2) in the setup of Accept–Reject methods where the rejected simulations can be recycled by integrating over the uniform variates that are used at the selection stage.

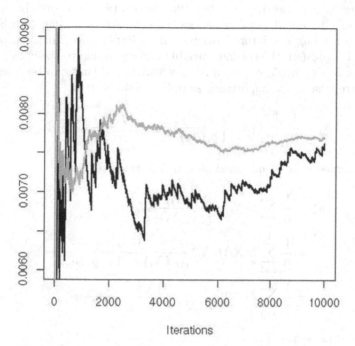

Iterations

Fig. 4.7. Convergence of the estimators of $\mathbb{E}[\exp(-X^2)]$, δ_n *(solid lines, darker)*, and $\delta_n^\star$ *(dots, lighter)* for $(\nu, \mu, \sigma) = (5, 3, 0.5)$. The final values are 0.00762 and 0.00769, respectively, for a true value equal to 0.00771.

Exercise 4.10 Consider an Accept–Reject method for the target density f based on the instrumental distribution g.

a. Show that the Accept–Reject sample $(X_1, \ldots, X_n)$ can be associated with two iid samples, $(U_1, \ldots, U_N)$ and $(Y_1, \ldots, Y_N)$, with corresponding distributions $\mathcal{U}_{[0,1]}$ and g; N is the stopping time associated with the acceptance of n variables Y_j.
b. Deduce that the corresponding estimator of $\mathbb{E}_f[h(X)]$ based on $(X_1, \ldots, X_n)$ can be written as

$$\delta_1 = \frac{1}{n} \sum_{i=1}^{n} h(X_i) = \frac{1}{n} \sum_{j=1}^{N} h(Y_j) \, \mathbb{I}_{U_j \le w_j},$$

with $w_j = f(Y_j)/Mg(Y_j)$.

Another generic setting for Rao–Blackwellization has already been illustrated in Chapter 3 with the notion of defensive sampling (see Section 3.3.3). When using a mixture distribution as importance function, $g(x) = \varrho g_1(x) + (1 - \varrho)g_2(x)$, the random variable corresponding to the choice of the component of the mixture is an auxiliary variable Y that can be integrated out. If we represent the importance sampling estimator by

$$\delta_n = \frac{1}{n} \sum_{i=1}^{n} h(X_i) \left\{ \frac{f(X_i)}{g_1(X_i)} \mathbb{I}_{Y_i=1} + \frac{f(X_i)}{g_2(X_i)} \mathbb{I}_{Y_i=2} \right\},$$

taking its expectation conditional on the X_i's leads to

$$\delta_n^\star = \frac{1}{n} \sum_{i=1}^{n} h(X_i) \mathbb{E}\left[\frac{f(X_i)}{g_{Y_i}(X_i)} \bigg| X_i \right]$$

$$= \frac{1}{n} \sum_{i=1}^{n} h(X_i) f(X_i) \frac{\varrho + 1 - \varrho}{\varrho g_1(X_i) + (1 - \varrho)g_2(X_i)}$$

$$= \frac{1}{n} \sum_{i=1}^{n} h(X_i) \frac{f(X_i)}{\varrho g_1(X_i) + (1 - \varrho)g_2(X_i)},$$

as suggested in Section 3.3.3.

The additional stabilization brought by setting the number of variables simulated from g_1 equal to ϱn and the number of variables simulated from g_2 equal to $(1 - \varrho)n$ is not, formally speaking, a Rao–Blackwellization, even though the resulting estimator is both unbiased and less variable than the estimator with a binomial $\mathcal{B}in(n, \varrho)$ number of generations from g_1.

When enough information is available on the density f, a specific implementation of Rao–Blackwellization is *stratified sampling*. The case occurs when the probabilities $\varrho_i = P_f(A_i)$ of the sets A_i of a given partition $\{A_1, \ldots, A_p\}$ of $\mathcal{X}$ are known. Since the integral $\int_{\mathcal{X}} h(x)f(x)\,\mathrm{d}x$ can be expressed as

$$\int_{\mathcal{X}} h(x)f(x)\,\mathrm{d}x = \sum_{i=1}^{p} \int_{A_i} h(x)f(x)\,\mathrm{d}x = \sum_{i=1}^{p} \varrho_i \int_{A_i} h(x)f_i(x)\,\mathrm{d}x = \sum_{i=1}^{p} \varrho_i \mathfrak{I}_i,$$

where the f_i's are the restrictions of f to the regions A_i $(i = 1, \ldots, p)$, from a Rao–Blackwellization point of view, this means that simulating $X \sim f$ is equivalent to simulating Y from $P(Y = i) = \varrho_i$ and then simulating $X|Y = i \sim f_i$, while the variability due to the simulation of Y can be integrated out.

In practice, samples of size n_i are generated from the f_i's to evaluate each integral $\mathfrak{I}_i$ separately by a regular estimator $\hat{\mathfrak{I}}_i$. The variance of the resulting estimator, $\varrho_1 \hat{\mathfrak{I}}_1 + \cdots + \varrho_p \hat{\mathfrak{I}}_p$, that is,

$$\sum_{i=1}^{p} \varrho_i^2 \frac{1}{n_i} \int_{A_i} (h(x) - \mathfrak{I}_i)^2 f_i(x)\,dx,$$

may be much smaller than the variance of the standard Monte Carlo estimator based on a sample of size $n = n_1 + \cdots + n_p$. The optimal choice of the n_i's in this respect is such that

$$(n_i^*)^2 \propto \varrho_i^2 \int_{A_i} (h(x) - \mathfrak{I}_i)^2 f_i(x)\,dx.$$

Thus, if the regions A_i can be chosen, the variance of the stratified estimator can be reduced by selecting A_i's with similar variance factors $\int_{A_i} (h(x) - \mathfrak{I}_i)^2 f_i(x)\,dx$. Of course, the number of cases where stratified sampling is possible is limited.

4.7 Acceleration methods

Similar to the deconditioning device above, there exist global *acceleration* strategies that are more or less independent of the simulation setup but try to exploit the output of the simulation in more efficient ways and should thus be implemented whenever possible. When seen as postprocessing devices, those methods can also be used to assess convergence by providing alternative estimators of the same quantity.

4.7.1 Correlated simulations

Although the usual simulation methods lead to iid samples, it may actually be preferable to generate samples of negatively or positively correlated variables when estimating an integral $\mathfrak{I}$, as they may reduce the variance of the corresponding estimator.

The first setting where the independence requirement may be less desirable corresponds to the comparison of two quantities that are close in value. If

$$(4.6) \qquad \mathfrak{I}_1 = \int g_1(x) f_1(x)dx \quad \text{and} \quad \mathfrak{I}_2 = \int g_2(x) f_2(x)dx$$

are two such quantities, where δ_1 estimates $\mathfrak{I}_1$ and δ_2 estimates $\mathfrak{I}_2$, independently of δ_1, the variance of $(\delta_1 - \delta_2)$ is then $\mathrm{var}(\delta_1) + \mathrm{var}(\delta_2)$, which may be too large to support a fine enough analysis on the difference $\mathfrak{I}_1 - \mathfrak{I}_2$. However, if δ_1 and δ_2 are positively correlated, the variance is reduced by a factor of $-2\,\mathrm{cov}(\delta_1, \delta_2)$, which may improve the analysis of the difference.

A convincing illustration of the improvement brought by correlated samples is the comparison of (regular) statistical estimators via simulation. Given a density $f(x|\theta)$ and a loss function $L(\delta, \theta)$, two estimators δ_1 and δ_2 are evaluated through their risk functions, $R(\delta_1, \theta) = \mathbb{E}[L(\delta_1, \theta)]$ and $R(\delta_2, \theta)$ (see, e.g., Robert, 2001, Chapter 2). In general, these risk functions are not available analytically, but they may be approximated, for instance, by a regular Monte Carlo method,

$$\hat{R}(\delta_1, \theta) = \frac{1}{n} \sum_{i=1}^{n} L(\delta_1(X_i), \theta), \qquad \hat{R}(\delta_2, \theta) = \frac{1}{n} \sum_{i=1}^{n} L(\delta_2(Y_i), \theta)$$

the X_i's and Y_i's being simulated from $f(\cdot|\theta)$. Positive correlation between $L(\delta_1(X_i), \theta)$ and $L(\delta_2(Y_i), \theta)$ then reduces the variability of the approximation of $R(\delta_1, \theta) - R(\delta_2, \theta)$.

Two general-purpose recommendations for conducting simulation-based comparisons of statistical procedures are that

(i). First, the *same sample* $(X_1, \ldots, X_n)$ should be used in the Monte Carlo approximations of $R(\delta_1, \theta)$ and $R(\delta_2, \theta)$. This repeated use of a single sample does not greatly jeopardize the convergence of the approximation, while it improves the precision of the estimated difference $R(\delta_1, \theta) - R(\delta_2, \theta)$, as shown by the comparison of the variances of $\hat{R}(\delta_1, \theta) - \hat{R}(\delta_2, \theta)$ and

$$\frac{1}{n} \sum_{i=1}^{n} \{L(\delta_1(X_i), \theta) - L(\delta_2(X_i), \theta)\}.$$

(ii). Second, *the same sample* should also be used for the comparison of risks *for every value of* θ (i.e., for the approximation of the entire risk function) in order to present a smoother representation of this function. Although this sounds like an absurd recommendation when the sample $(X_1, \ldots, X_n)$ is usually generated from a distribution depending on θ, it is often the case that the same uniform sample can be used for the generation of the X_i's for every value of θ. Also, in many settings, there exists a transformation M_θ on $\mathcal{X}$ such that if $X^0 \sim f(X|\theta_0)$, $M_\theta X^0 \sim f(X|\theta)$. A single sample $(X_1^0, \ldots, X_n^0)$ from $f(X|\theta_0)$ is then sufficient to produce a sample from $f(X|\theta)$ using the transform M_θ. Finally, the implementation of this principle is obvious when using importance sampling.

The second point is somewhat tangential for the theme of this section; however, it brings significant improvement in the practical implementation of Monte Carlo methods.

The variance reduction associated with the conservation of the underlying uniform sample is obvious in the graphs of the resulting risk functions, which then miss the irregular peaks of graphs obtained with independent samples and allow an easier comparison of estimators.

Example 4.7. In the case $X \sim \mathcal{N}_p(\theta, I_p)$, the transform M_θ mentioned in the remark above obviously is the location shift $M_\theta X = X + \theta - \theta_0$. When studying *positive-part James–Stein estimators*

$$\delta_a(x) = \left(1 - \frac{a}{\|x\|^2}\right)^+ x, \qquad 0 \le a \le 2(p-2)$$

(see Robert, 2001, Chapter 2, for a motivation), the comparison of the squared error risk of the δ_a's is more easily conducted by simulation, based on a single normal sample, as follows:

```
> nor=matrix(rnorm(Nsim*p),nrow=p)
> risk=matrix(0,ncol=150,nrow=10)
> a=seq(1,2*(p-2),le=10)
> the=sqrt(seq(0,4*p,le=150)/p)
> for (j in 1:150){
+    nornor=apply((nor+rep(the[j],p))^2,2,sum)
+    for (i in 1:10){
+    for (t in 1:Nsim)
+       risk[i,j]=risk[i,j]+sum((rep(the[j],p)-
+       max(1-a[i]/nornor[t],0)*(nor[,t]+rep(the[j],p)))^2)
+    }}
> risk=risk/Nsim
```

Figure 4.8 illustrates this comparison in the case $p = 5$. Note that the upper curve, which corresponds to the estimated risk of δ_a for $a = 2*(p-2)$, exceeds the value $p = 5$ for larger θ's since it ends up at 5.12, while the theory guarantees that the true risk does not. This is the consequence of using a Monte Carlo approximation, obviously, and running more simulations would lead to a value closer to $p = 5$, as you can check. A derivation of the empirical error of risk[i,j] shows that the precision is ± 0.23, which means that the confidence interval on the risk of $\delta_{2(p-2)}$ includes values that are lower than 5. ◀

In a more general setup, creating a strong enough correlation between δ_1 and δ_2 is rarely that simple, and the quest for correlation can result in an increase in the conception and simulation burdens, which may even have a negative overall effect on the efficiency of the analysis. Indeed, to use the same uniform sample for the generation of variables distributed from f_1 and f_2 in (4.6) is only possible when there exists a simple transformation from f_1 to f_2 such as location-scale. For instance, if f_1 or f_2 must be simulated by Accept–Reject methods, the use of a random number of uniform variables prevents the use of a common sample.

4.7.2 Antithetic variables

The method of *antithetic variables* is based on the same idea that higher efficiency can be obtained through correlation. Given two samples $(X_1,\ldots, X_n)$ and $(Y_1,\ldots,Y_n)$ from f used for the estimation of

$$\Im = \int_{\mathbb{R}} h(x)f(x)\,dx,$$

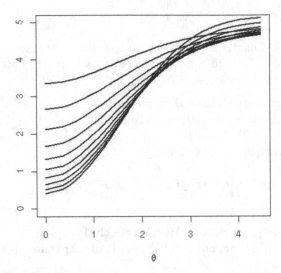

Fig. 4.8. Approximate squared error risks of truncated James–Stein estimators for a normal distribution $\mathcal{N}_5(\theta, I_5)$, as a function of $\|\theta\|$, based on $n = 10^3$ simulations.

the estimator

$$(4.7) \qquad \frac{1}{2n} \sum_{i=1}^{n} [h(X_i) + h(Y_i)]$$

is more efficient than an estimator based on an iid sample of size $2n$ if the variables $h(X_i)$ and $h(Y_i)$ are *negatively correlated*. In this setting, the Y_i's are called the *antithetic variables*. Obviously, creating negative correlation for an arbitrary transform h is not always possible, even when X_i and Y_i are negatively correlated. A solution proposed in Rubinstein (1981) is to use uniform variables U_i to generate the X_i's as $X_i = F^-(U_i)$ and the Y_i's as $Y_i = F^-(1-U_i)$. This idea can be generalized toward a systematic coverage of the unit interval that is closely related to quasi-random schemes (see, e.g., Robert and Casella, 2004, Section 2.6.2). Given a first sample of $X_i = F^-(U_i)$'s, 2^q replicated samples can be constructed by considering the 2^q first dyadic symmetries of the U_i's, obtained by switching the q first bits of the binary decompositions of the U_i's.

Exercise 4.11 Show that if $H = h \circ F^-$, $U \sim \mathcal{U}(0,1)$, $X = F^-(U)$, and $Y = F^-(1-U)$, then $h(X)$ and $h(Y)$ are negatively correlated when H is a monotone function.

Example 4.8. (Continuation of Example 4.1) Based on the original simulation experiment

```
> uref=runif(10^4)
> x=h(uref)
> estx=cumsum(x)/(1:10^4)
```

the dyadic symmetries can be produced by

```
> resid=uref%%2^(-q)
> simx=matrix(resid,ncol=2^q,nrow=10^4)
> simx[,2^(q-1)+1:2^1]=2^(-q)-simx[,2^(q-1)+1:2^1]
> for (i in 1:2^q) simx[,i]=simx[,i]+(i-1)*2^(-q)
> xsym=h(simx)
> estint=cumsum(apply(xsym,1,mean))/(1:10^4)
```

through a systematic exploration of all the terms in the unit interval with the same terminal bits. The impact of averaging over those values is clear from Figure 4.9. Using the symmetries over the first four bits produces a much more stable sequence, while using the first eight bits gives the correct answer 9.9652 (obtained by integrate) almost from the start, even though $2^8 = 256$ is much smaller than 10^4. ◄

As can be extrapolated from this example, the improvement brought by the dyadic averaging is of a numerical nature; i.e., it relates to a perspective of numerical integration rather than Monte Carlo integration. Indeed, if the number q of bits goes to infinity (and Example 4.8 showed that $q = 8$ is already close to infinity), the random aspect of the estimator vanishes. As a consequence, settings where this technique is likely to have an impact are those where numerical integration applies.

The technique only applies for simulations that are direct transforms of uniform variables, thus excluding the Accept–Reject methods where the impact of the negative correlation is most likely diluted. More general group actions can nonetheless be considered, as in Evans and Swartz (2000) and Kong et al. (2003), where the authors replace the standard average by an average (over i) of the average of the $h(gX_i)$ (over the transformations g).[3] An immediate application of this averaging principle is to replicate the available sample using random permutations over the level sets of the density; i.e., replicating X_i into

$$Y_i \sim \mathcal{U}\left(\{y;\, f(y) = f(X_i)\}\right).$$

Example 4.9. (Continuation of Example 4.5) If we see the numerator of the posterior mean as an expectation under the normal $\mathcal{N}(x, 1)$ distribution,

---

[3] Strictly speaking, this is not antithetic sampling since the averaging is over the dominating measure rather than on the distribution itself.

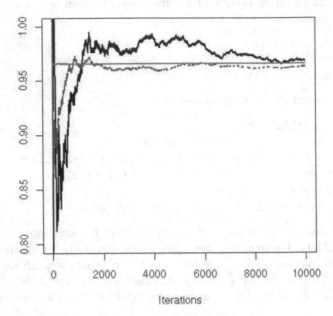

Fig. 4.9. Impact of the dyadic average over the approximation of the integral of h studied in Example 4.1 for 2^4 *(grey, dotted)* and 2^8 replicas *(light grey straight line)* when compared with the convergence of the original sequence *(dark full line)*.

the symmetry is with respect to x; i.e., the sample of θ_i's is to be replicated into $2x - \theta_i$'s, while, if the reference distribution is the Cauchy distribution, the sample of θ_i's is duplicated into the sample of $-\theta_i$'s. ◄

⚡ In the few situations where simulation from this uniform distribution can be conducted, the improvement brought by averaging via Monte Carlo over level sets is not always very significant.

4.7.3 Control variates

In some particular cases, there exist functions h_0 whose mean under f is known. For instance, if f is symmetric around μ or if μ is the median of f, the mean of $h_0(X) = \mathbb{I}_{X \geq \mu}$ is $1/2$. The technique of *control variates* takes advantage of this additional piece of information to reduce the variance of a Monte Carlo estimator of $\mathfrak{I} = \int h(x) f(x) \mathrm{d}x$ in the following way. If δ_1 is an estimator of $\mathfrak{I}$ and δ_3 an unbiased estimator of $\mathbb{E}_f[h_0(X)]$, consider the weighted estimator

$$\delta_2 = \delta_1 + \beta(\delta_3 - \mathbb{E}_f[h_0(X)]).$$

The estimators δ_1 and δ_2 have the same mean and

$$\operatorname{var}(\delta_2) = \operatorname{var}(\delta_1) + \beta^2 \operatorname{var}(\delta_3) + 2\beta \operatorname{cov}(\delta_1, \delta_3).$$

For the optimal choice

$$\beta^* = -\frac{\operatorname{cov}(\delta_1, \delta_3)}{\operatorname{var}(\delta_3)},$$

we have

$$\operatorname{var}(\delta_2) = (1 - \rho_{13}^2)\operatorname{var}(\delta_1),$$

ρ_{13}^2 being the correlation coefficient between δ_1 and δ_3, so the control variate strategy will result in a decreased variance for δ_2. In particular, if

$$\delta_1 = \frac{1}{n}\sum_{i=1}^{n} h(X_i) \qquad \text{and} \qquad \delta_3 = \frac{1}{n}\sum_{i=1}^{n} h_0(X_i),$$

the control variate estimator is

$$\delta_2 = \frac{1}{n}\sum_{i=1}^{n} \left(h(X_i) + \beta^* \{h_0(X_i) - \mathbb{E}_f[h_0(X)]\} \right),$$

with $\beta^* = -\operatorname{cov}(h(X), h_0(X))/\operatorname{var}(h_0(X))$.

When p control variates are available, $h_{01}(X_i), \ldots, h_{0p}(X_i)$ say, the extension is immediate, the control variate estimator then being

$$\delta_2 = \frac{1}{n}\sum_{i=1}^{n} \left(h(X_i) + \sum_{j=1}^{p} \beta_j^* \{h_{0j}(X_i) - \mathbb{E}_f[h_{0j}(X)]\} \right)$$

and the coefficient vector $\beta^* = (\beta_1^*, \ldots, \beta_p^*)$ being derived as the orthogonal projection

$$\beta^* = -\operatorname{var}_f(\mathbf{h}_0(X))^{-1} \operatorname{cov}_f(h(X), \mathbf{h}_0(X))$$

for $\mathbf{h}_0(x) = (h_{01}(x), \ldots, h_{0p}(x))$.

Exercise 4.12 Show that if $f(x|\theta)$ is the density of interest, parameterized by $\theta \in \mathbb{R}$, the function $h(x, \theta) = \partial \log f(x|\theta)/\partial\theta$ is always a control variate with mean zero. Discuss the relevance of this result when f is known up to a constant (as a density in x).

In practice, β^* obviously is unavailable but can be approximated by a simple regression of the $h(x_i)$'s over the $h_0(x_i)$'s (or a multiple regression of the $h(x_i)$'s over the $h_{0j}(x_i)$'s).

Example 4.10. (Continuation of Example 4.9) Since the normal distribution used in the representation of the numerator of the posterior mean is centered at x, all odd moments of $(x-\theta)$ have zero mean. It is therefore possible to conduct a regression of the simulated $\theta_i/(1+\theta_i^2)$'s on the control variates $(x-\theta_i)^{2k+1}$ in order to improve the approximation of

$$\mathfrak{I} = \int \frac{\theta}{1+\theta^2} \frac{\exp -(x-\theta)^2/2}{\sqrt{2\pi}} \, d\theta.$$

A rudimentary implementation of the control variate technique is as follows:

```
> thet=rnorm(10^3,mean=x)
> delt=thet/(1+thet^2)
> moms=delta=c()
> for (i in 1:5){
+    moms=rbind(moms,(thet-x)^(2*i-1))
+    reg=lm(delt~t(moms)-1)$coef
+    delta=rbind(delta,as.vector(delt-reg%*%moms))
+    }
> plot(cumsum(delt)/(1:10^3),ty="l",lwd=2,lty=2)
> for (i in 1:5) lines(cumsum(delta[i,])/(1:10^3),lwd=2)
```

It should stop adding variates when the regression fit does not improve and when the regressors become nonsignificant. In Table 4.1, the residual error stops decreasing for $k = 3$, where the regression summary is

```
    Coefficients:
             Estimate Std. Error t value Pr(>|t|)
    moms1   -0.110128   0.027401  -4.019 6.28e-05 ***
    moms2    0.004218   0.016815   0.251   0.802
    moms3    0.002693   0.001986   1.356   0.176


    Residual standard error: 0.3486 on 997 degrees of freedom
    Multiple R-Squared: 0.05643,    Adjusted R-squared: 0.05359
    F-statistic: 19.87 on 3 and 997 DF,  p-value: 1.621e-12
```

This indicates that the third odd moment (i.e., the fifth moment) is not contributing significantly to the regression, while, for $k = 2$, the regression summary is

```
    Coefficients:
             Estimate Std. Error t value Pr(>|t|)
    moms1   -0.137687   0.018380  -7.491 1.50e-13 ***
    moms2    0.025994   0.004976   5.224 2.14e-07 ***
```

The graphical comparison in Figure 4.10 of the estimators thus induced shows no difference from $k = 3$ onward and very similar variabilities for $k = 2$ and $k = 3$. The improvement upon the original graph of estimates is quite apparent. ◀

Table 4.1. Comparisons of the first regression coefficient and the regression fit parameters against the number k of odd moments used as control variates.

k	1	2	3	4	5
β_1	−.06	−.113	−.110	−.105	−.108
R^2	.0279	.0528	.0536	.0527	.0517
$\hat{\sigma}$	.353	.349	.349	.349	.349

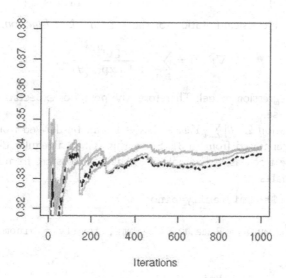

Iterations

Fig. 4.10. Impact of the control variates $(x - \theta)^{2k-1}$ over the approximation of the integral $\mathfrak{I}$ for $k = 1, \dots, 5$. The original sequence of estimators is the dotted one *(dark)*, while the control variate corrections are increasing with k at iteration 500 *(grey)*. The graphs for $k = 3, 4, 5$ are indistinguishable.

The technique of control variates thus appears to be manageable only in very specific situations where the control function h must be available. There exists, however, a class of Bayesian models where some expectations always are available, in the setting of conjugate priors and exponential families (see, e.g., Robert, 2001, Chapter 3).

Example 4.11. Consider modeling the Pima.tr dataset—introduced in Section 1.5 and already used in Example 3.10—using a logistic regression model, $(1 \leq i \leq m)$

$$P(Y_i = 1|x_i) = \exp(x_i^t\theta)/\{1 + \exp(x_i^t\theta)\}.$$

Under a flat prior on θ, the posterior distribution on the regression coefficient has density proportional to

$$(4.8) \qquad \exp\left(\theta^t \sum_i Y_i x_i\right) \prod_{i=1}^{m} \{1 + \exp(x_i^t \theta)\}^{-1}.$$

As for all distributions from an exponential family, we have that

$$\mathbb{E}_\theta\left[\sum_{i=1}^{m} Y_i x_i\right] = m \nabla \psi(\theta) \quad \text{and} \quad \mathbb{E}^\pi\left[\nabla \psi(\theta) \middle| \sum_{i=1}^{m} Y_i x_i\right] = \frac{\sum_i Y_i x_i}{m},$$

where $\psi(\theta)$ is the classical notation for the *log-cumulant function*, such that

$$m \nabla \psi(\theta) = \sum_{i=1}^{m} \frac{\exp(x_i^t \theta)}{1 + \exp(x_i^t \theta)} x_i$$

in the logistic regression model. Therefore, the posterior expectation of $\nabla \psi(\theta)$ is known in this case.

The expectation $\mathbb{E}^\pi[\theta| \sum_i Y_i x_i + \zeta, \lambda + 1]$ can be derived from variables θ_j $(1 \leq j \leq n)$ generated from (4.8). Since this is not a regular distribution, we follow the same track as in Example 3.10, namely to start from the maximum likelihood estimates

```
> glm(Pima.tr$t~bmi,family=binomial)
```

```
Call:  glm(formula = Pima.tr$t ~ bmi, family = binomial)
```

```
Coefficients:
(Intercept)         bmi
   -0.7249       0.1048
```

to construct a scaled normal proposal as in

```
> sim=cbind(rnorm(10^4,m=-.72,sd=.55),rnorm(10^4,m=.1,sd=.2))
> weit=apply(sim,1,like)/(dnorm(sim[,1],m=-.72,sd=.55)*
+            dnorm(sim[,2],m=.1,sd=.2))
```

The efficiency is again rather poor, with a normalized effective sample size of 7% and a perplexity of 9.5%.

A control variate version of

$$\delta_1 = \sum_{j=1}^{n} \omega_j \theta_j \middle/ \sum_{j=1}^{n} \omega_j$$

is thus available via the weighted regression of the θ_j's upon the

$$\sum_{i=1}^{m} \frac{\exp x_i^t \theta_j}{1 + \exp x_i^t \theta_j} x_i,$$

which can be readily obtained as

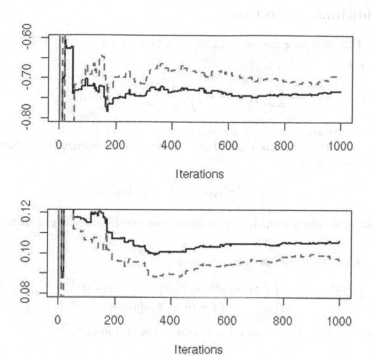

Fig. 4.11. Impact of the log-cumulant control variates over the posterior estimations of the logistic regression coefficients for the Pima.tr dataset. The original sequence of estimators over the first 10^3 iterations is the full one *(dark)*, while the control variate correction is the dotted curve *(grey)*. The upper graph corresponds to the intercept, with maximum likelihood estimate equal to -0.725, and the lower graph to the coefficient of the body mass index, with maximum likelihood estimate equal to 0.105.

```
> vari1=(1/(1+exp(-sim[,1]-sim[,2]*bmi)))-
+ sum((Pima.tr$t=="Yes"))/length(Pima.tr$bmi)
> vari2=(bmi/(1+exp(-sim[,1]-sim[,2]*bmi)))-
+     sum(bmi[Pima.tr$t=="Yes"])/length(Pima.tr$bmi)
> resim=sample(1:Nsim,Nsim,rep=T,pro=weit)
> reg=as.vector(lm(sim[resim,1]~t(rbind(vari1[resim],
+                         vari2[resim]))-1)$coef)
```

since resampling avoids using weighted regression. The impact on the estimation of the coefficients of θ is, however, quite limited, as is shown in Figure 4.11. ◄

4.8 Additional exercises

Exercise 4.13 Following the results obtained in Exercise 4.2:

a. Deduce that

$$(4.9) \qquad \frac{1}{n_2} \sum_{i=1}^{n_2} \tilde{\pi}_1(\theta_{2i}) \alpha(\theta_{2i}) \Bigg/ \frac{1}{n_1} \sum_{i=1}^{n_1} \tilde{\pi}_2(\theta_{1i}) \alpha(\theta_{1i}) \,,$$

with $\theta_{1i} \sim \pi_1$ and $\theta_{2i} \sim \pi_2$, is a convergent estimator of ϱ.

b. Show that part b covers as a special case the Newton and Raftery (1994) harmonic mean representation

$$\varrho = \mathbb{E}^{\pi_2} [\tilde{\pi}_2(\theta)^{-1}] \Bigg/ \mathbb{E}^{\pi_1} [\tilde{\pi}_1(\theta)^{-1}]$$

if both expectations exist. Give a sufficient condition for those expectations to be finite.

Exercise 4.14 Using the following approximation

$$\frac{\mathrm{Var}(\hat{\varrho})}{\varrho^2} \approx \frac{1}{n_1 n_2} \left\{ \frac{\int \pi_1(\theta)\pi_2(\theta)[n_1\pi_1(\theta) + n_2\pi_2(\theta)]\alpha(\theta)^2 \, \mathrm{d}\theta}{\left(\int \pi_1(\theta)\pi_2(\theta)\alpha(\theta) \, \mathrm{d}\theta \right)^2} - 1 \right\}$$

to the variance of (4.9), show that the optimal choice of α in (4.9) is

$$\alpha(\theta) \propto 1 \Bigg/ n_1 \pi_1(\theta) + n_2 \pi_2(\theta) \,.$$

The optimal solution exhibited in Exercise 4.14 above can also be interpreted as a defensive mixture version, as in (3.8), while the harmonic mean solution is generally worthless because of its infinite variance.

Exercise 4.15 For each of the following cases, generate random variables X_i and Y_i and compare the empirical average and Rao–Blackwellized estimator of $\mathbb{E}_f(X)$ and $\mathrm{var}_f(X)$:

a. $X|y \sim \mathcal{P}(y)$, $Y \sim \mathcal{G}a(a,b)$ (X is negative binomial);
b. $X|y \sim \mathcal{N}(0, y)$, $Y \sim \mathcal{G}a(a,b)$ (X is a generalized t);
c. $X|y \sim Bin(y)$, $Y \sim Be(a,b)$ (X is beta-binomial).

Exercise 4.16 Given an Accept–Reject algorithm based on (f, g, ρ), we denote by

$$b(y_j) = \frac{(1-\rho)f(y_j)}{g(y_j) - \rho f(y_j)}$$

the importance sampling weight of the rejected variables $(Y_1, \dots, Y_t)$ and by $(X_1, \dots, X_n)$ the accepted variables.

a. Show that the estimator

$$\delta_1 = \frac{n}{n+t} \delta^{\mathrm{AR}} + \frac{t}{n+t} \delta_0,$$

with $\delta_0 = \frac{1}{t} \sum_{j=1}^t b(Y_j) h(Y_j)$ and $\delta^{\mathrm{AR}} = \frac{1}{n} \sum_{i=1}^n h(X_i)$, does not uniformly dominate δ^{AR}. (*Hint:* Consider the constant functions.)

b. Show that

$$\delta_{2w} = \frac{n}{n+t} \delta^{AR} + \frac{t}{n+t} \sum_{j=1}^{t} \frac{b(Y_j)}{S_t} h(Y_j)$$

is asymptotically equivalent to δ_1 in terms of bias and variance.

c. Deduce a condition for δ_{2w} to asymptotically dominate δ^{AR} (see Robert and Casella, 2004, Section 4.6.2).

Exercise 4.17 In conjunction with Exercise 4.6, show that, if we denote by Σ_k^{-1} the inverse matrix corresponding to $(\bar{X}_1, \bar{X}_2, \dots, \bar{X}_k)$, to get Σ_{k+1}^{-1} we only have to change one element and then add one row and one column to Σ_k^{-1}.

Exercise 4.18 A naïve way to implement the antithetic variable scheme is to use both U and $(1-U)$ in an inversion simulation. Examine empirically whether this method leads to variance reduction for the distributions (i) $f_1(x) = 1/\pi(1+x^2)$, (ii) $f_2(x) = \frac{1}{2}e^{-|x|}$, (iii) $f_3(x) = e^{-x}\mathbb{I}_{x>0}$, (iv) $f_4(x) = \frac{2}{\pi\sqrt{3}}\left(1+x^2/3\right)^{-2}$, and (v) $f_5(x) = 2x^{-3}\mathbb{I}_{x>1}$. Examine variance reductions of the mean, second moment, median, and 75th percentile.

To calculate the weights for the Rao–Blackwellized estimator of Section 4.6, it is necessary to derive properties of the distribution of the random variables in the Accept–Reject algorithm. The following problem is a rather straightforward exercise in distribution theory and is only made complicated by the stopping rule of the Accept–Reject algorithm.

Exercise 4.19 This problem looks at the performance of a *termwise* Rao–Blackwellized estimator. Casella and Robert (1998) established that such an estimator does not sacrifice much performance over the full Rao–Blackwellized estimator of Exercise 4.10. Given a sample $(Y_1, \dots, Y_N)$ produced by an Accept–Reject algorithm to accept n values based on (f, g, M):

a. Show that

$$\frac{1}{N} \sum_{i=1}^{N} \mathbb{E}[\mathbb{I}_{U_i \le \omega_i} | Y_i] h(Y_i) = \frac{1}{N-n} \left(h(Y_N) + \sum_{i=1}^{N-1} b(Y_i) h(Y_i) \right)$$

with

$$b(Y_i) = \left(1 + \frac{n(g(Y_i) - \rho f(Y_i))}{(N-n-1)(1-\rho)f(Y_i)} \right)^{-1}.$$

b. If $S_N = \sum_1^{N-1} b(Y_i)$, show that

$$\delta = \frac{1}{N-n} \left(h(Y_N) + \frac{N-n-1}{S_n} \sum_{i=1}^{N-1} b(Y_i) h(Y_i) \right)$$

asymptotically dominates the usual Monte Carlo approximation, conditional on the number of rejected variables n under quadratic loss. (*Hint:* Show that the sum of the weights S_N can be replaced by $(N-n-1)$ in δ and assume $\mathbb{E}_f[h(X)] = 0$.)

Exercise 4.20 Following from Exercise 4.10,

a. Conclude that a reduction in the variance of δ_1 can be obtained by integrating out the U_i's, as in the estimator

$$\delta_2 = \frac{1}{n} \sum_{j=1}^{N} \mathbb{E}[\mathbb{I}_{U_j \leq w_j} | N, Y_1, \ldots, Y_N] \, h(Y_j) = \frac{1}{n} \sum_{i=1}^{N} \rho_i h(Y_i).$$

b. Show that, for $i = 1, \ldots, k - 1$, ρ_i satisfies

$$\rho_i = P(U_i \leq w_i | N = k, Y_1, \ldots, Y_n)$$
$$= w_i \frac{\sum_{(i_1, \ldots, i_{n-2})} \prod_{j=1}^{n-2} w_{ij} \prod_{j=n-1}^{k-2}(1 - w_{ij})}{\sum_{(i_1, \ldots, i_{n-1})} \prod_{j=1}^{n-1} w_{ij} \prod_{j=n}^{k-1}(1 - w_{ij})},$$

while $\rho_k = 1$. The numerator sum is over all subsets of $\{1, \ldots, i-1, i+1, \ldots, k-1\}$ of size $n - 2$, and the denominator sum is over all subsets of size $n - 1$.

Exercise 4.21 The control variate scheme can be adapted to the Accept–Reject algorithm. When $Y_1, \ldots, Y_N$ is the sample produced by an Accept–Reject algorithm based on g aiming at t acceptances, let m denote the density

$$m(y) = \frac{t - 1}{n - 1} f(y) + \frac{n - t}{n - 1} \frac{g(y) - \rho f(y)}{1 - \rho}$$

when $N = n$ and $\rho = \dfrac{1}{M}$.

a. Show that m is the marginal density of the Y_i's conditional upon $N = n$, and deduce that

$$\mathfrak{I} = \int h(x) f(x) \mathrm{d}x = \mathbb{E}_N \left[\mathbb{E}^m \left[\frac{h(Y_i) f(Y_i)}{m(Y_i)} \Big| N \right] \right].$$

b. Show that, for any function $c(\cdot)$ with a closed-form expectation $\mathbb{E}[c(Y)]$ and for any constant β,

$$\mathfrak{I} = \beta \mathbb{E}[c(Y)] + \mathbb{E} \left[\frac{h(Y) f(Y)}{m(Y)} - \beta c(Y) \right].$$

Deduce that

$$\hat{\mathfrak{I}} = \frac{h(Y_N)}{N} + \frac{1}{N} \sum_{i=1}^{N-1} \left\{ \frac{h(Y_i) f(Y_i)}{m(Y_i)} - \beta \left(c(Y_i) - \mathbb{E}[c(Y)] \right) \right\}$$

is a control variate estimator of $\mathfrak{I}$.

c. Setting $d(y) = h(y) f(y)/m(y)$, show that the optimal choice of β is

$$\beta^* = \mathrm{cov}[d(Y), c(Y)]/\mathrm{var}[c(Y)].$$

d. Examine choices of c for which the optimal β can be constructed and thus where the control variate method applies.

(*Note:* Strawderman, 1996, suggests estimating β using $\hat{\beta}$, the regression coefficient of $d(Y_i)$ on $c(Y_i)$, $i = 1, 2, \ldots, n - 1$.)

5

Monte Carlo Optimization

"He invented a game that allowed players to predict the outcome?"

Susanna Gregory

To Kill or Cure

Reader's guide

This chapter is the equivalent for optimization problems of what Chapter 3 is for integration problems. We distinguish between two separate uses of computer-generated random variables to solve optimization problems. The first use, as seen in Section 5.3, is to produce stochastic search techniques to reach the maximum (or minimum) of a function, devising random exploration techniques on the surface of this function that avoid being trapped in local maxima (or minima) and are sufficiently attracted by the global maximum (or minimum). The second use, described in Section 5.4, is closer to Chapter 3 in that simulation is used to approximate the function to be optimized.

C.P. Robert, G. Casella, *Introducing Monte Carlo Methods with R*, Use R,
DOI 10.1007/978-1-4419-1576-4_5, © Springer Science+Business Media, LLC 2010

5.1 Introduction

Optimization problems can mostly be seen as one of two kinds: We either need to find the extrema of a function $h(\theta)$ over a domain Θ or find the solution(s) to an implicit equation $g(\theta) = 0$ over a domain Θ. Both problems are exchangeable to some extent in that the second one is a minimization problem for a function like $h(\theta) = g^2(\theta)$ (in dimension one), while the first one is equivalent to solving $\partial h(\theta)/\partial \theta = 0$ (assuming the function h can be differentiated). Therefore, we only focus on the maximization problem

$$(5.1) \qquad\qquad\qquad \max_{\theta \in \Theta} h(\theta)$$

since a minimization problem can be handled as a maximization problem when substituting $-h$ or $1/h$ for h.[1]

Similar to the problem of integration treated in Chapter 3, the optimization problem (5.1) can be processed by either numerical or stochastic means. In the numerical perspective, performance is highly dependent on the analytical properties of the target function (such as convexity, boundedness, and smoothness), while those properties of h play a lesser role in simulation-based approaches. Therefore, if h is too complex to allow an analytic study or if the domain Θ is too irregular, the method of choice is rather the stochastic approach.

The entire chapter addresses the issue of finding extrema using stochastic techniques, but let us point out here that optimization problems are generically harder to solve than integration problems. That is, the former are much more local than the latter, at least from a computational point of view (since, from a mathematical viewpoint, the information about the extrema of a function is as global as that about its integrals). In other words, it is harder to pinpoint a single extreme point in a domain than the average of a regular function over the same domain. Note that, in accord with the remainder of the book, this chapter deals with optimization problems that relate to statistics, with domains that are almost always continuous. This means that hard combinatoric problems leading to optimization over a finite (if large) set such as the *traveling salesman problem* (see, e.g., Spall, 2003, Robert and Casella, 2004) are not considered here.

---

[1] While the choice of the transform is innocuous from a mathematical point of view, since the argument of the minimum does not change, its impact on the approximation method used to find this argument is far from innocuous and the selection of the transform should be considered carefully.

5.2 Numerical optimization methods

In R, there are several embedded functions to solve optimization problems. The simplest one is `optimize` (or `optimise`), which processes one-dimensional targets.

Example 5.1. When considering maximizing the likelihood of a Cauchy $\mathcal{C}(\theta, 1)$ sample,

$$\ell(\theta|x_1, \ldots, x_n) = \prod_{i=1}^{n} \frac{1}{1 + (x_i - \theta)^2},$$

the sequence of maxima (i.e., of the MLEs) is converging to $\theta^* = 0$ when n goes to ∞. This is reflected by Figure 5.1 *(left)*, which corresponds to the code

```
> xm=rcauchy(500)
> f=function(y){-sum(log(1+(x-y)^2))}
> for (i in 1:500){
+   x=xm[1:i]
+   mi=optimise(f,interval=c(-10,10),max=T)$max}
```

where the log-likelihood is maximized sequentially as the sample increases. However, when looking directly at the likelihood, using `optimise` eventually produces a diverging sequence since the likelihood gets too small around $n = 300$ observations, even though the sequences of MLEs are the same up to this point for both functions. When we replace the smooth likelihood with a perturbed version, as in

```
> f=function(y){-sin(y*100)^2-sum(log(1+(x-y)^2))}
```

the `optimise` function gets much less stable, as demonstrated in Figure 5.1 *(right)*, since the two sequences of MLEs (corresponding to the log-likelihood and the likelihood) are no longer identical.   ◀

⚡ The output of `optimise` is a list with components "minimum" (or "maximum") and "objective". Rather unfortunately, in contrast with mathematical conventions, "minimum" gives the *location* of the minimum of the function (that is, the *argument* of the minimum), while "objective" gives the value of the minimum of the function (that is, the minimum itself)!

Similarly, `nlm` is a generic R function that searches for the minimum of a function based on the Newton–Raphson method—that is, based on the recurrence relation

$$\theta_{i+1} = \theta_i - \left[\frac{\partial^2 h}{\partial\theta\partial\theta^{\mathrm{T}}}(\theta_i) \right]^{-1} \frac{\partial h}{\partial\theta}(\theta_i)$$

—where the matrix of the second derivatives is called the *Hessian* and the vector of the first derivatives the *gradient* (sometimes denoted ∇h). This method

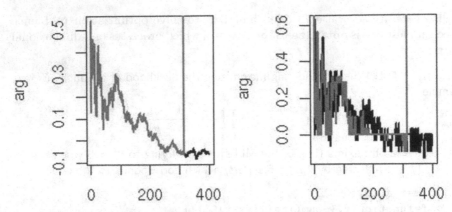

Fig. 5.1. *(left)* Sequence of MLEs corresponding to 500 simulations from a Cauchy $\mathcal{C}(0, 10)$ distribution obtained by applying optimize to the log-likelihood and the likelihood *(in lighter colors)*; *(right)* the same sequences when using a perturbed likelihood.

is perfect when h is quadratic but may also deteriorate when h is highly non-linear, and it obviously does not work when the domain Θ is irregular. It also obviously depends on the starting point θ_0 when h has several minima.

Example 5.2. The likelihood associated with the mixture model

$$(5.2) \qquad \qquad \frac{1}{4}\mathcal{N}(\mu_1, 1) + \frac{3}{4}\mathcal{N}(\mu_2, 1)$$

is bimodal, as seen in Figure 5.2 for a simulated sample of 400 observations from this mixture with $\mu_1 = 0$ and $\mu_2 = 2.5$, actually produced by

```
> da=rbind(rnorm(10^2),2.5+rnorm(3*10^2))
> like=function(mu){
+    sum(log((.25*dnorm(da-mu[1])+.75*dnorm(da-mu[2]))))}
```

and by applying the R function contour to a grid of points where the log-likelihood function like is computed. When using nlm, the modes are obtained within a few iterations, depending on the starting points, and the intermediate values of the Newton–Raphson sequence can be plotted by

```
> sta=c(1,1)
> mmu=sta
> for (i in 1:(nlm(like,sta)$it))
+    mmu=rbind(mmu,nlm(like,sta,iter=i)$est)
> lines(mmu,pch=19,lwd=2)
```

where the function like has been redefined as its inverse to account for the fact that nlm produces a local minimum. Note that some starting points produce warnings:

`NA/Inf replaced by maximum positive value in: nlm(like, sta)`

meaning that the (numerical approximation to the) Hessian is not invertible at
the current value. The sequences represented in Figure 5.2 all end up in one of the
two modes, but with highly nonlinear patterns. For instance, the starting point
$(-1, -1)$ corresponds to a very steep gradient and thus bypasses the main mode
$(-0.68, 1.98)$ to end up at the secondary one (lower in likelihood). Although all
represented sequences do converge, starting farther away from the modes may
produce divergent sequences.   ◄

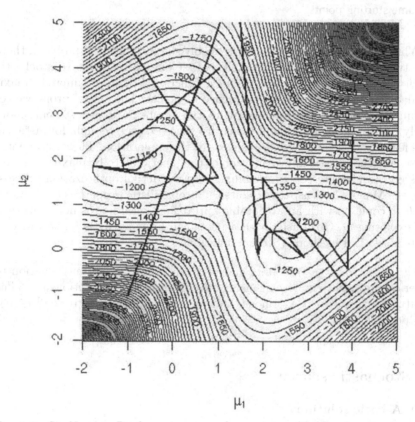

Fig. 5.2. Six Newton–Raphson sequences for a mixture likelihood ending in one of
two modes depending on the starting point based on a sample of 400 observations
from the normal mixture (5.2) with $\mu_1 = 0$ and $\mu_2 = 2.5$.

Exercise 5.1 Write an R code that truly produces a sample of 400 observations
from (5.2) instead of setting the normal subsample sizes to 100 and 300 as above.
Compare the shapes of the corresponding log-likelihoods in both cases.

The function nlm is a numerical method to determine the minimum of its first argument f. While it is not absolutely exact and depends on its starting value p, we stress that nlm is deterministic. Therefore, repeatedly using nlm with the same starting value p will always produce the same Newton–Raphson sequence. On the other hand, if the target h is modified by a monotonic transform—that thus leaves the true mode unchanged— the Newton–Raphson sequence will differ and may end up diverging in some cases.

Exercise 5.2 Using 1/like as the target of nlm in the setting of Example 5.2, compare the Newton–Raphson sequences with those produced using -like and the same starting points.

When using simulation techniques to solve an optimization problem, there exist two different points of entry. The first one corresponds to stochastic search or exploration methods, where a function h is approximately maximized by considering a random sequence of points. The actual properties of the function play a lesser role here, with the Monte Carlo aspect being more closely tied to the exploration of the entire space Θ, even though, for instance, some features of h can be used to speed up the exploration. The second entry is based on a stochastic approximation of the function h to be maximized, and this step can also be seen as preliminary to the actual optimization process. Here, the Monte Carlo aspect exploits the probabilistic properties of the function h to come up with an acceptable approximation $\hat{h}$ and is not concerned with exploring Θ. Obviously, both aspects can be merged, as for instance in Monte Carlo EM methods (Section 5.4.4).

This is obviously a very brief and necessarily incomplete introduction to numerical optimization methods using only the most basic functions in R. The literature in this area is enormous, and we refer for instance to Spall (2003) for a possible entry.

5.3 Stochastic search

5.3.1 A basic solution

A natural if rudimentary way of using simulation to obtain an approximation to the solution of (5.1) is to simulate points over Θ according to an arbitrary distribution f positive everywhere on Θ until a sufficiently high value of $h(\theta)$ is observed. This solution may be very inefficient if f is not chosen in connection with h but, given an infinite number of simulations and some regularity requirements on the problem, including the compactness of the domain Θ, it is bound to converge (see, e.g., Spall, 2003, Theorem 2.1). For instance, if Θ is bounded, we can simulate from a uniform distribution on Θ, $u_1, \ldots, u_m \sim \mathcal{U}_\Theta$, and use $h_m^* = \max(h(u_1), \ldots, h(u_m))$ as an approximation to the solution of (5.1).

Example 5.3. Recall the simple and regular but highly variable function $h(x) = [\cos(50x) + \sin(20x)]^2$ defined on $[0, 1]$, first seen in Example 3.3. A call to optimise provides an identification of the maximum at $x^* = 0.379$ with a value of $h(x^*) = 3.8325$. If we want to assess the variability of a uniform sampler, we can use multiple uniform sequences as in

```
> rangom=h(matrix(runif(10^6),ncol=10^3))
> monitor=t(apply(rangom,1,cummax))
> plot(monitor[1,],type="l",col="white")
> polygon(c(1:10^3,10^3:1),c(apply(monitor,2,max),
+          rev(apply(monitor,2,min))),col="grey")
> abline(h=optimise(h,int=c(0,1),max=T)$ob)
```

whose result is shown in Figure 5.3. While the starting value $h(u_1)$ of the sequence is highly variable, the range reduces very quickly and, after 10^3 iterations, the worst sequence among the 10^3 parallel sequences is within 0.24 of the true maximum. ◄

Using a uniform distribution over the domain Θ is only relevant when Θ has a regular shape. Otherwise, it is more efficient to simulate from a uniform distribution over a domain containing Θ and to drop the simulations outside Θ.

Exercise 5.3 When Θ is defined on $\mathbb{R}^2$ by the constraint

$$x^2(1 + \sin(3y)\cos(8x)) + y^2(2 + \cos(5x)\cos(8y)) \leq 1,$$

propose a simple uniform simulation on a larger domain and evaluate the performance of this method via the average number of rejected points.

Obviously, this blind solution—blind in the sense that it does not take h into account—quickly gets impractical as the dimension or the complexity of the problem increases. For instance, in a Bayesian setting, when the size n of an iid sample $x_1, \ldots, x_n$ from $f(x|\theta)$ grows, the associated posterior distribution $\pi(\theta|x_1, \ldots, x_n)$ gets more and more concentrated around its mode, which is thus more and more difficult to approximate this way. This, however, is not always the case, as the following example shows.

Example 5.4. (Continuation of Example 5.1) Using the same Cauchy model as in Example 5.1, we can monitor the discrepancy between the solutions provided by optimise and by uniform sampling over $[-5, 5]$ as the sample size n increases from $n = 1$ to $n = 5001$. Figure 5.4 *(top)* plots the value of the true argument θ^* (as provided by optimise) against the value of the uniform

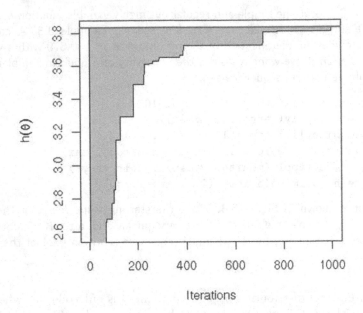

Fig. 5.3. Range of 10^3 sequences of successive maxima found by random uniform sampling over 10^3 iterations. The true maximum value is identified by the grey line on top of the graph.

sample with the highest likelihood and does not show a significant increase of the error near zero since there are more points in this neighborhood. Similarly, Figure 5.4 *(bottom)* shows that the relative error between the true maximum and the approximation on the uniform sample does not exhibit an increasing trend.   ◄

It is therefore more fruitful to design the simulation experiment in (close) connection with h as well as with the domain Θ. Intuitively, it makes sense to increase the probability of simulating in regions where h is large and to decrease this probability in regions where it is small. This means creating a probability distribution that is related with h in a nonlinear way but with identical or close to identical modes. Obviously, if h is positive and

$$\int_\Theta h(\theta)\, d\theta < +\infty,$$

the resolution of (5.1) amounts to finding the *modes* of the density proportional to h. More generally, any density H sharing maxima with h is a potentially interesting choice, for instance

$$H(\theta) \propto \exp(h(\theta)/T)$$

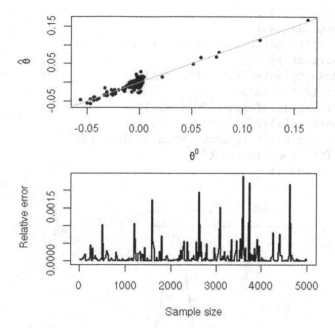

Fig. 5.4. Comparison of a numerical and a stochastic maximization of a Cauchy likelihood in terms of the sample size via *(top)* respective locations of the numerical and stochastic evaluations of the arguments, plotted along the diagonal; *(bottom)* relative error of the stochastic evaluation against the numerical evaluation as a function of the sample size.

for any $T > 0$ such that $\exp(h(\theta)/T)$ is integrable. The parameter T, called the *temperature*, is free to calibrate and can be chosen toward accelerating convergence or avoiding local maxima, as we will discuss later. The problem (5.1) then being expressed in statistical terms, it is natural to generate a sample $(\theta_1, \ldots, \theta_m)$ from H and to apply a standard mode estimation method on this sample (or to simply compare the $h(\theta_i)$'s). In some cases, it may be more practical to decompose $H(\theta)$ into $H(\theta) = H_1(\theta)H_2(\theta)$ and to simulate from H_1 only.

In statistical applications, h is most often a likelihood and thus simulating from h is identical with simulating from the posterior distribution associated with the flat prior.

Example 5.5. (Continuation of Example 5.1) The likelihood is clearly interpretable as a posterior distribution on θ. However, this is not a standard distribution, and we need a handy substitute! Given that the product of the Cauchy densities amounts to the inverse of a polynomial (in θ) of degree $2n$, we can pick a t distribution with $(n-1)/2$ degrees of freedom, with mean the empirical median—since the Cauchy mean is notoriously unstable—and scale the interquartile range. Comparing the true posterior with the approximation via

```
> cau=rcauchy(10^2)
> mcau=median(cau)
> rcau=diff(quantile(cau,c(.25,.75)))
> f=function(x){
+    z=dcauchy(outer(x,cau,FUN="-"))
+    apply(z,1,mean)}
> fcst=integrate(f,from=-20,to=20)
> ft=function(x){f(x)/fcst}
> g=function(x){dt((x-mcau)/rcau,df=49)/rcau}
> curve(ft,from=-10,to=10)
> curve(g,add=T)
```

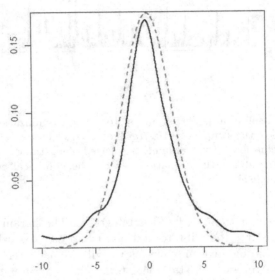

Fig. 5.5. Comparison of the posterior density *(in black)* of a Cauchy location parameter based on 101 observations and a t approximation *(dotted lines)*.

shows that the approximation is acceptable, as shown in Figure 5.5. (Note the use of the R function outer in the function f, already discussed in Example 3.6. This function allows us to apply the function f to a vector, if not a matrix, and thus to use curve.) As expected, using the t approximation produces evaluations of the maximum likelihood that converge faster (in the number of simulations), as can be seen through the following Monte Carlo experiment (which is easy to reproduce on your own).

```
> unisan=matrix(f(runif(5*10^4,-5,5)),ncol=500)
> causan=matrix(f(rt(5*10^4,df=dft)*rcau+mcau),ncol=500)
> unimax=apply(unisan,2,cummax)[10:10^2,]
```

```
> caumax=apply(causan,2,cummax)[10:10^2,]
> plot(caumax[,1],col="white",ylim=c(.8,1)*max(causan))
> polygon(c(10:10^2,10^2:10),c(apply(unimax,1,max),
+   rev(apply(unimax,1,min))),col="grey")
> polygon(c(10:10^2,10^2:10),c(apply(caumax,1,max),
+   rev(apply(caumax,1,min))),col="wheat")
```

In line with the idea above of decreasing the *temperature* to freeze the simulations in higher and higher values of the target function h, it would be possible to repeat the experiment with a smaller range, except that our t approximation does not preserve the mode of h, so this strategy is not advisable in the present case. ◄

Exercise 5.4 Given the function $h(x)$ defined in Example 5.3, deduce from the bound
$$h(x) \leq 2 * (|\cos(50x)| + |\sin(20x)|)$$
a practical way to simulate from a density proportional to h. Compare the variability in the evaluation of $\max h(x)$ based on 10^3 simulations from h with the variability in the same evaluation based on 10^3 uniform simulations.

While this solution is completely natural and formally provides a converging method, its implementation is far from obvious. Finding a density H that both shares modes with h and is easy to simulate is quite a challenge when h is a complex function, and the solutions we will produce in Sections 5.3.2 and 5.3.3 are actually inferring on h locally (that is, in a neighborhood of the current simulations from H), in a spirit almost identical to the MCMC algorithms of Chapters 6 and 7.

Example 5.6. Consider minimizing the (artificially constructed) function in $\mathbb{R}^2$
$$h(x,y) = (x\sin(20y) + y\sin(20x))^2 \cosh(\sin(10x)x)$$
$$+ (x\cos(10y) - y\sin(10x))^2 \cosh(\cos(20y)y),$$
whose global minimum is 0, attained at $(x,y) = (0,0)$. Since this function has many local minima, as shown by Figure 5.6, obtained via

```
> h=function(x,y){(x*sin(20*y)+y*sin(20*x))^2*cosh(sin(10*x)
+          *x)+(x*cos(10*y)-y*sin(10*x))^2*cosh(cos(20*y)*y)}
> x=y=seq(-3,3,le=435)              #defines a grid for persp
> z=outer(x,y,h)
> par(bg="wheat",mar=c(1,1,1,1))   #bg stands for background
> persp(x,y,z,theta=155,phi=30,col="green4",
+          ltheta=-120,shade=.75,border=NA,box=FALSE)
```

it does not satisfy the conditions under which standard minimization methods are guaranteed to provide the global minimum. On the other hand, the distribution on $\mathbb{R}^2$ with density proportional to $\exp(-h(x,y))$ can be simulated, even though this is not a standard distribution, using either an Accept–Reject algorithm based on the uniform distribution (since h is positive)—which is then defeating the purpose of simulating from h rather than from the uniform distribution over Θ!—or more advanced MCMC techniques introduced later in Chapter 6.   ◀

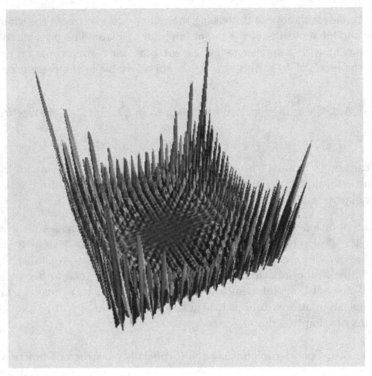

Fig. 5.6. Representation via persp of the function $h(x, y)$ of Example 5.6 on $[-3, 3]^2$.

5.3.2 Stochastic gradient methods

Given that generating direct simulations from the target function H defined in the previous section is often a major difficulty, a different stochastic approach to the maximization of h is to explore the surface of h in a local manner (that is, by defining a sequence $\{\theta_j\}_j$ by moving from θ_j to θ_{j+1} in a dependent step) rather than independently as in the basic stochastic search algorithm. The dependence of θ_{j+1} on θ_j is often chosen to be linear, in the sense that it is represented as

$$(5.3) \qquad\qquad \theta_{j+1} = \theta_j + \epsilon_j \,,$$

where ϵ_j is the local perturbation of the current value. In mathematical terms, this makes the sequence $\{\theta_j\}_j$ a Markov chain. While there is a connection between those methods and the MCMC algorithms (Chapters 6 and 7), the Markov property, however, is less important in the present setting simply because the mathematics justifying the convergence to a global maximum is too advanced to be considered here (see, e.g., Hàjek, 1988 or Haario and Sacksman, 1991).

When considering the implementation of the local update idea in (5.3), the perturbation ϵ_j can be simulated from an arbitrary distribution such as a $\mathcal{N}_p(0, \sigma^2 I_p)$ distribution if $\Theta \subset \mathbb{R}^p$. However, given that we are specifically searching for the maximum of h, using some information about h in constructing the distribution of the perturbation is bound to increase the efficiency of the method. In particular, it makes sense to favor moves increasing in h over moves decreasing in h, even though the latter should not be impossible if you want to avoid local maxima. A natural approach is to use the gradient of h, ∇h, if available.

In numerical optimization, the *gradient method* is a deterministic numerical approach to the optimization problem (5.1) related to the Newton–Raphson method already introduced in Section 5.2. It produces a sequence $\{\theta_j\}$ defined by

$$(5.4) \qquad \theta_{j+1} = \theta_j + \alpha_j \nabla h(\theta_j), \qquad \alpha_j > 0,$$

that converges to the exact solution of (5.1), θ^*, when the domain $\Theta \subset \mathbb{R}^d$ and the function $(-h)$ are both convex—thus assuming there exists a single maximum—and for various choices of the decreasing sequence $\{\alpha_j\}$ (see Thisted, 1988). For less regular problems, the gradient sequence is most likely to get stuck in a local extremum of the function h.

Stochastic gradient methods take advantage of this method to build the perturbation in (5.3). For instance, the *finite-difference* proposal is to build a numerical substitute to the true gradient

$$\nabla h(\theta_j) \approx \frac{h(\theta_j + \beta_j \zeta_j) - h(\theta_j + \beta_j \zeta_j)}{2\beta_j} \zeta_j = \frac{\Delta h(\theta_j, \beta_j \zeta_j)}{2\beta_j} \zeta_j,$$

where (β_j) is a second decreasing sequence and ζ_j is uniformly distributed over the unit sphere $\|\zeta\| = 1$. In contrast to the deterministic approach, the update

$$(5.5) \qquad \theta_{j+1} = \theta_j + \frac{\alpha_j}{2\beta_j} \Delta h(\theta_j, \beta_j \zeta_j) \zeta_j$$

does not proceed along the steepest slope of h in θ_j since each time it picks the direction at random, but this property is generally a *plus* in the sense that it may avoid being trapped in local maxima or in saddlepoints of h.

Whether or not $\{\theta_j\}$ defined by (5.5) does converge to the argument θ^* of (5.1) will highly depend on the choice of the sequences $\{\alpha_j\}$ and $\{\beta_j\}$. For instance, α_j needs to decrease slowly enough to 0 for the corresponding series $\sum_j \alpha_j$ to diverge, while β_j must decrease even more slowly for the series $\sum_j (\alpha_j/\beta_j)^2$ to converge (Spall, 2003, Chapter 6).

Example 5.7. (Continuation of Example 5.6) We apply the iterative construction (5.5) to the multimodal function $h(x, y)$ with different sequences of α_j's and β_j's to check for their impact. A natural stopping rule for the algorithm is to check for stabilization in the sequence θ_j, leading to the R implementation

```
> start=c(.65,.8)
> theta=matrix(start,ncol=2)
> diff=iter=1
> while (diff>10^-5){
+    zeta=rnorm(2)
+    zeta=zeta/sqrt(t(zeta)%*%zeta)
+    grad=alpha[iter]*zeta*(h(theta[iter,]+beta[iter]*zeta)-
+         h(theta[iter,]-beta[iter]*zeta))/beta[iter]
+    theta=rbind(theta,theta[iter,]+grad)
+    dif=sqrt(t(grad)%*%grad)
+    iter=iter+1}
```

where the sequences of α_j's and β_j's have to be inserted. When running this R code, we actually had to include an inner safety loop

```
> scale=sqrt(t(grad)%*%grad)
> while (scale>1){
+    zeta=rnorm(2);zeta=zeta/sqrt(t(zeta)%*%zeta)
+    grad=alpha[iter]*zeta*(h(theta[iter,]+beta[iter]*zeta)-
+         h(theta[iter,]-beta[iter]*zeta))/beta[iter]
+    scale=sqrt(t(grad)%*%grad)}
```

to protect against diverging evaluations of the gradient grad that occur from time to time and lead the program to abort.

The different sequences we tested are

Scenario	1	2	3	4
α_j	$1/\log(j+1)$	$1/100\log(j+1)$	$1/(j+1)$	$1/(j+1)$
β_j	$1/\log(j+1)^{.1}$	$1/\log(j+1)^{.1}$	$1/(j+1)^{.5}$	$1/(j+1)^{.1}$

Note that, in each case, the decrease in the second sequence is much slower than for the first sequence. When using faster sequences (β_j), you should check that the method does not necessarily converge to the global minimum.

Figure 5.7 illustrates on single runs (you are encouraged to duplicate those runs) that, depending on the speed of convergence of (α_j), the global minimum is correctly approximated or not. In the case of scenario 1, where both α_j and β_j decrease very slowly, it appears that the perturbation lacks the energy necessary

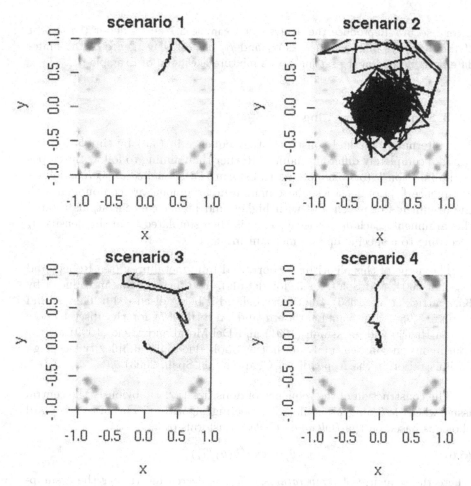

Fig. 5.7. Single realizations of stochastic gradient paths for four different choices of the sequences (α_j) and (β_j) with the same starting point $(0.65, 0.8)$: Scenario 1 corresponds to $(\alpha_j, \beta_j) = (1/100\log(j+1), 1/\log(j+1)^{-1})$, scenario 2 corresponds to $(\alpha_j, \beta_j) = (1/\log(j+1), 1/\log(j+1)^{-1})$, scenario 3 corresponds to $(\alpha_j, \beta_j) = (1/(j+1), 1/(j+1)^{.5})$, and scenario 4 corresponds to $(\alpha_j, \beta_j) = (1/(j+1), 1/(j+1)^{-1})$. The function h to be minimized is defined in Example 5.6 and the minimum of h is achieved at the central point $(0, 0)$.

to reach the global minimum, while multiplying (α_j) by 100 shows a sequence θ_j that does not stabilize quickly enough to remain at the global minimum (as can be checked via theta[iter,]). For the geometric decreases of scenarios 3 and 4, the difference in the power involved in β_j does not significantly impact the ability to uncover the true minimum of h. ◄

Exercise 5.5 Reproduce the analysis of Example 5.7 above about the impact of the dynamics of the sequences α_j and β_j on the convergence of the finite-difference method in the setting of the mixture likelihood of Example 5.2.

5.3.3 Simulated annealing

This alternative method constructs the sequence in (5.3) by simulating the ϵ_j's in a completely different manner. Rather than aiming to follow the slopes of the function h (or a monotonic transform H), *simulated annealing* defines a sequence $\{\pi_t\}$ of densities whose maximum arguments are confounded with the arguments of $\max h$ and with higher and higher concentrations around this argument. Each θ_t in the sequence is then simulated from the density π_t according to a specific update mechanism.

The name of this algorithm is borrowed from metallurgy (see Robert and Casella, 2004, and Spall, 2003, for details). The method was introduced by Kirkpatrick et al. (1983), the theory behind it being discussed in Geman and Geman (1984) and Van Laarhoven and Aarts (1987) for the discrete case and in Haario and Sacksman (1991) and Del Moral and Miclo (1999) for the continuous one, but certainly not in this book, due to its complexity (see, e.g., an illustration in the Appendix of Chapter 8 in Spall, 2003).

The construction of the sequence of densities $\{\pi_t\}$ is obviously the central issue when designing a simulated annealing algorithm. The most standard choice is based on the *Boltzman–Gibbs* transforms of h,

$$(5.6) \qquad \pi_t(\theta) \propto \exp(h(\theta)/T_t),$$

where the sequence of *temperatures*, $\{T_t\}$, is decreasing (under the assumption that the right-hand side of (5.6) is integrable). It is indeed clear that, as T_t decreases toward 0, values simulated from π_t become concentrated in a narrower and narrower neighborhood of the maximum (or maxima) of h. The additional feature of simulated annealing, when compared with the basic simulation approach of Section 5.3.1, is that this technique does not simulate an entire sample from π_t at each iteration t but accelerates convergence by simulating a *single realization* from π_t. It thus updates both the sequence *and* the target distribution at each iteration t.

The choice (5.6) is a generic solution to concentrate (in t) the distribution π_t around the maxima of an arbitrary function h, but other possibilities are available in specific settings. For instance, when maximizing a likelihood $\ell(\theta|x)$, the pseudo-posterior distributions $\pi_t(\theta|x) \propto \ell(\theta|x)^{m_t}\pi_0(\theta)$, associated with a nondecreasing integer sequence $\{m_t\}$ and an arbitrary prior π_0, enjoy the same property. This alternative is obviously more intuitive from a statistical point of view since it amounts to using a regular likelihood with a sample replicated $(m_t - 1)$ times, and it has been introduced in Robert (1993) under the name of *prior feedback* (see also Robert and Casella, 2004, Section 5.2.4) and studied for latent variable models in Doucet et al. (2002) under the acronym of *SAME* (state augmentation for marginal estimation). (This algorithm has been rediscovered under various names by Gaetan and Yao, 2003, Lele et al., 2007 or Jacquier et al., 2007.) Simulation algorithms used in Bayesian analysis such as those presented in Chapters 6 and 7 can obviously be recycled for this purpose.

Exercise 5.6 For the Cauchy likelihood of Example 5.1, based on a simulated sample of size $n = 100$, show that the pseudo-posterior distribution $\pi_m(\theta|x) \propto \ell(\theta|x)^m$ is defined for any integer $m > 0$. Using `integrate` to properly normalize π_m, show graphically how π_m concentrates as m increases.

Two practical issues that hinder the implementation of this otherwise attractive algorithm are (a) the simulation from π_t and (b) the selection of the temperature sequence (or *schedule*) $\{T_t\}$. While the second item is very problem-dependent, the first item allows a generic solution, related to the Metropolis–Hastings algorithm, first proposed by Metropolis et al. (1953) and detailed in Chapter 6. The update from θ_t to θ_{t+1} is indeed based on the Metropolis–Hastings step: ζ is generated from a distribution with symmetric density g, and the new value θ_{t+1} is generated as

$$\theta_{t+1} = \begin{cases} \theta_t + \zeta & \text{with probability } \rho = \exp(\Delta h/T_t) \wedge 1, \\ \theta_t & \text{with probability } 1 - \rho, \end{cases}$$

where $\Delta h = h(\zeta + \theta_t) - h(\theta_t)$.

Since we are not interested in proving convergence results for this algorithm, we will postpone until the next chapter the proof that this transition preserves the distribution π_t (provided it exists) and will focus on its connection with the maximization problem. Instead of looking for a deterministic or stochastic gradient, the algorithm proposes a symmetric perturbation of the current value, $\theta_t + \zeta$. If the perturbation increases h (i.e., if $h(\theta_t + \zeta) \geq h(\theta_t)$), the new value is automatically accepted. On the other hand, if $h(\theta_t + \zeta) < h(\theta_t)$, this move may still be accepted with probability $\rho > 0$. Otherwise, a new perturbation $\theta_t + \zeta$ is created and tested (from a maximization viewpoint, the fact that $\theta_{t+1} = \theta_t$ does not really matter, except in the determination of the stopping rule). By allowing random moves that may see h decrease, the simulated annealing method can explore multimodal functions and escape the attraction of local modes as opposed to deterministic (and to some extent stochastic) gradient methods.

The algorithmic rendering of the simulated annealing is thus

Algorithm 2 Simulated Annealing
At iteration t,
   1. Simulate $\zeta \sim g(\zeta)$;
   2. Accept $\theta_{t+1} = \theta_t + \zeta$ with probability
     $\rho_t = \exp\{\Delta h_t / T_t\} \wedge 1$;
     take $\theta_{t+1} = \theta_t$ otherwise.

---

the density g being symmetric (around 0) but otherwise almost arbitrary.

An R version of this algorithm is associated with a random generator from g, randg, as in Algorithm 1,

```
> theta=rep(theta0,Nsim)
> hcur=h(theta0)
> xis=randg(Nsim)
> for (t in 2:Nsim){
+    prop=theta[t-1]+xis[t]
+    hprop=h(prop)
+    if (Temp[t]*log(runif(1))<hprop-hcur){
+    theta[t]=prop
+    hcur=hprop
+    }else{
+       theta[t]=theta[t-1]}}
```

where the temperature sequence Temp needs to be defined by the user.

Obviously, the performance of the algorithm will depend on the choice of g. For instance, if g corresponds to perturbations with a large scale, the moves will most often be rejected because Δh_t will be negative most of the time. On the other hand, if the scale of g is small, the sequence $\{\theta_t\}$ will have difficulties in exploring several modes and will most likely end up being stuck at the mode it started with, thus cancelling the appeal of the method. As will be discussed in Chapter 6, a proper scaling of g should correspond to an acceptance rate between .2 and .6.

Example 5.8. (Continuation of Example 5.3) For the simple function from Example 5.3, $h(x) = [\cos(50x) + \sin(20x)]^2$, we can compare the impact of using different temperature schedules on the performance of the simulated annealing sequences. Note that, besides setting a temperature sequence, we also need to set a scale value (or sequence) for the distribution g of the perturbations as well as a stopping rule. Since the domain is $[0,1]$, we use a uniform $\mathcal{U}(-\rho, \rho)$ distribution for g and our stopping rule is that the algorithm will stop when the observed maximum of h has not changed in the second half of the sequence $\{x_t\}$.

An R rendering of this simulated annealing algorithm is

```
> x=runif(1)
> hval=hcur=h(x)
```

```
> diff=iter=1
> while (diff>10^(-4)){
+   prop=x[iter]+runif(1,-1,1)*scale
+   if ((prop>1)||(prop<0)||
        (log(runif(1))*temp[iter]>h(prop)-hcur))
+                prop=x[iter]
+   x=c(x,prop)
+   hcur=h(prop)
+   hval=c(hval,hcur)
+   if ((iter>10)&&(length(unique(x[(iter/2):iter]))>1))
+       diff=max(hval)-max(hval[1:(iter/2)])
+   iter=iter+1}
```

The constraint involving unique is cancelling the stopping rule when no perturbation has been accepted in the second half of the iterations, meaning that the scale may then be inappropriate. (Note that the updates of temp and scale need to be included in the loop.)

For a scale defined by $\sqrt{T_t}$ and a temperature decrease in $1/\log(1+t)$, the sequence almost always ends up at a value close to the true maximum. Similarly, a scale defined by $5\sqrt{T_t}$ and a temperature decrease in $1/(1+t)^2$ leads almost certainly to the global maximum, as shown on Figure 5.8 (where the last example was obtained after several runs). Decreasing the scale by a factor of ten has a clear and negative impact on the performance of the algorithm. ◀

You can, in particular, check by testing the code above that the faster T_t decreases to 0, the less likely the simulated annealing sequence is to leave the current mode.

While there exist theoretical results about temperature schedules that guarantee convergence of the simulated annealing algorithm, they have little practical value because they depend on calibration constants that are problem-related. The general recommendation for the temperature decrease is that it should be logarithmic, as in $T_i = \Gamma/\log i$, rather than geometric, $T_i = \alpha^i T_0$, even though the former induces very slow convergence patterns. Adaptive strategies that update temperature and scale after learning episodes of several iterations that evaluate acceptance rates and maximum increase are thus recommended, even though their validation is mostly empirical.

The fact that approximate methods are necessary for optimization problems in *finite* state spaces may sound rather artificial and unnecessary, but the spaces involved in some modelings can be huge. For instance, in the *traveling salesman problem*, comparing all possible travels between n consecutive towns is of order $O(n!)$, which amounts to 10^{158} possible sequences for $n = 100$ towns. In the mixture setting of Example 5.2, the number of partitions of a sample of 400 observations into two groups is 2^{400} (i.e., more than 10^{120}). In genetics, the analysis of DNA sequences may involve 600,000 bases (A, C,

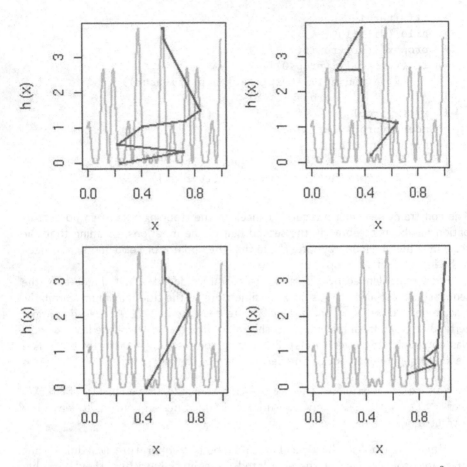

Fig. 5.8. Realizations of four simulated annealing sequences for $T_t = 1/(t+1)^2$ and $\rho = 5\sqrt{T_t}$ over the graph of the function h *(grey)*. Note that the points represented on the graph of h correspond to successive accepted values in Algorithm 2 and do not reflect the number of iterations.

G, or T), which corresponds to state-spaces of size $4^{600,000}$ if we consider all possible combinations.

Example 5.9. Using the same normal mixture likelihood as in Example 5.2, we can implement the simulated annealing algorithm for this example using for instance the following R function:

```
SA=function(x){
temp=scale=iter=dif=factor=1
the=matrix(x,ncol=2)
curlike=hval=like(x)
```

```
while (dif>10^(-4)){
  prop=the[iter,]+rnorm(2)*scale[iter]
  if ((max(-prop)>2)||(max(prop)>5)||
     (temp[iter]*log(runif(1))>-like(prop)+curlike))
         prop=the[iter,]
  curlike=like(prop);hval=c(hval,curlike);the=rbind(the,prop)
  iter=iter+1;temp=c(temp,1/10*log(iter+1))
  ace=length(unique(the[(iter/2):iter,1]))
  if (ace==1) factor=factor/10
  if (2*ace>iter) factor=factor*10
  scale=c(scale,max(2,factor*sqrt(temp[iter])))
  dif=(iter<100)+(ace<2)+(max(hval)-max(hval[1:(iter/2)]))
  }
  list(theta=the,like=hval,ite=iter)
}
```

As shown in Figure 5.9, the outcome is quite satisfactory. Most sequences end up in a close neighborhood of the maximum. It is also of interest to notice that the sequences are quite insensitive to the proximity of a given mode in that they most often visit the other mode before converging. (You should check that other realizations of the sequences may visit the second mode of the likelihood.) ◀

Exercise 5.7 In the setting of Example 5.9, build a Monte Carlo experiment that evaluates the frequency of visits to both modes of the likelihood for different temperature schedules.

Example 5.10. (Continuation of Example 5.6) We can also apply Algorithm 2 to find the minimum of the function h of Example 5.6. The perturbation is chosen to be Gaussian

```
> prop=the[iter,]+scale[iter]*rnorm(2)
```

as in the previous examples, and the scale is based on the current temperature,

```
> scale=min(.1,5*factor*sqrt(temp[iter]))
```

where factor depends on the acceptance rate of the algorithm, as in Example 5.9. As illustrated by Figure 5.10, the results change with the rate of decrease of the temperature T_i, both in the minima obtained (this will vary depending on the simulation, as you should check) and in the pattern of exploration of the valleys of h on both sides of the central zone. (Note that the heading of the four graphs was obtained by

```
> title(main=paste("min",format(-max(hval),dig=3),sep=" "))
```

using format to control the number of digits.) ◀

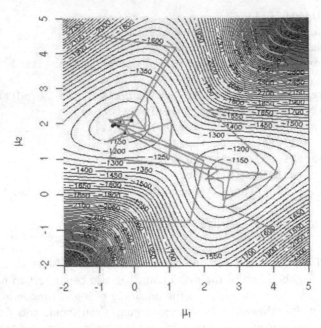

Fig. 5.9. Six simulated annealing sequences for a temperature schedule $T_t = 1/\log(1+t)$ based on a sample of 400 observations from the normal mixture (5.2) with $\mu_1 = 0$ and $\mu_2 = 2.5$.

5.4 Stochastic approximation

We now consider methods that work more directly with the objective function h rather than being concerned with fast exploration of the domain Θ for its optimization. Informally speaking, simulation is used in these methods to approximate the function h. In particular, when compared with the previous section, the use of those methods results in an additional level of error due to this approximation of h.

Although this may sound like an overspecialized problem at this stage, we will see in Section 5.4.2 that many statistical problems can be expressed in terms of an objective function h written as $\mathbb{E}[H(x, Z)]$. This is the setting of so-called *missing-data models*, which arise in many realistic setups. Moreover, note that artificial extensions (or *demarginalization*), which use this representation, are only computational devices and do not invalidate the overall inference. Before launching into their description, we cover the specific issue of maximizing an approximation of h.

5.4.1 Optimizing Monte Carlo approximations

If $h(x)$ can be written as $\mathbb{E}[H(x, Z)]$ but is not directly computable, a natural Monte Carlo approximation of h is

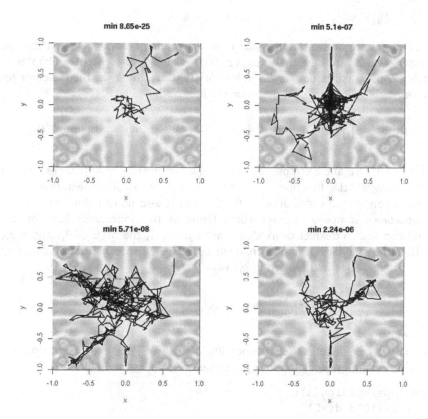

Fig. 5.10. Simulated annealing sequences for four temperature schedules: $T_t = (.95)^t$, $T_t = 1/10(t+1)$, $T_t = 1/\log(1+t)$, and $T_t = 1/10\sqrt{\log(1+t)}$, and starting point $(0.65, 0.8)$, aimed at minimizing the function h of Example 5.6. The light dot on top of the sequence corresponds to the final stage of the sequence $\{\theta_t\}$ and not necessarily the minimizer of h.

$$(5.7) \qquad \hat{h}(x) = \frac{1}{m} \sum_{i=1}^{m} H(x, z_i),$$

where the Z_i's are generated from the conditional distribution $f(z|x)$. This approximation yields a convergent estimator of $h(x)$ for every value of x (that is, it provides a *pointwise convergent estimator*), but its use in optimization setups is not recommended because, since the sample of Z_i's changes with every value of x, using an iterative optimization algorithm over the x's will result in an unstable sequence of evaluations of h and thus in a rather noisy resolution to $\arg\max h(x)$.

Example 5.11. In the Bayesian analysis of a simple probit model, where a binary variable $Y \in \{0, 1\}$ is distributed conditionally on a covariate X as

$$P_\theta(Y = 1|X = x) = 1 - P_\theta(Y = 0|X = x) = \Phi(\theta_0 + \theta_1 x),$$

it is of interest to derive the marginal posterior mode of, say, θ_0. For illustration purposes, we will here the Pima.tr dataset already presented in Chapter 1, X being chosen as the body mass index variate, bmi, and Y as the indicator for diabetes, type. The marginal posterior mode is obtained as

$$\arg\max_{\theta_0} \int \prod_{i=1}^n \Phi(\theta_0 + \theta_1 x_n)^{y_i} \Phi(-\theta_0 - \theta_1 x_n)^{1-y_i} \, d\theta_1 = \arg\max_{\theta_0} h(\theta_0)$$

for a flat prior on θ and a sample $(x_1, \ldots, x_n)$. Given the lack of analytic expression for this integral, the function h is then only defined as an expectation in θ_1. Since the conditional distribution of θ_1 given θ_0 is also nonstandard, we opt for an importance sampling approximation. Using as the importance function a t distribution with 5 degrees of freedom, mean $\mu = 0.1$, the MLE of θ_1, and scale $\sigma = 0.03$ the standard deviation provided by glm, we can construct a sample of θ_1^m $(m = 1, \ldots, M)$ and replace $h(\theta_0)$ with

$$\widehat{h}_0(\theta_0) = \frac{1}{M} \sum_{m=1}^M \prod_{i=1}^n \Phi(\theta_0 + \theta_1^m x_n)^{y_i} \Phi(-\theta_0 - \theta_1^m x_n)^{1-y_i} t_5(\theta_1^m; \mu, \sigma)^{-1},$$

where $t_5(\theta_1; \mu, \sigma)$ denotes the corresponding t density. Plotting this approximation of h with t samples simulated for each value of θ_0 using the R function

```
margap=function(a){
  b=rt(10^3,df=5)
  dtb=dt(b,5,log=T)
  b=b*.1+.1
  themar=0
  for (i in 1:10^3)
    themar=themar+exp(like(a,b[i])-dtb[i])
  themar/10^3
  }
```

(with like being defined as the probit likelihood as detailed in the remark below) shows how variable the approximation can be. Figure 5.11 *(top)* illustrates this variation both for one realization of $\widehat{h}$ and for a range of its variation based on 100 copies. It is obvious that the maximization of the represented $\widehat{h}$ function is not to be trusted as an approximation to the maximization of h.

In comparison, if we use the *same* t sample for all values of θ_0, we obtain a much smoother function, as shown by the central panel of Figure 5.11. While the range is naturally the same as in the top panel, the smoothness of the resulting $\widehat{h}$ function allows for a more trustworthy approximation. The bottom panel compares the averages of the approximations $\widehat{h}$ over the 100 replications for both approaches, showing no visible difference, which indicates that the corresponding 10^5 simulations are sufficient to produce a stable approximation of h.   ◀

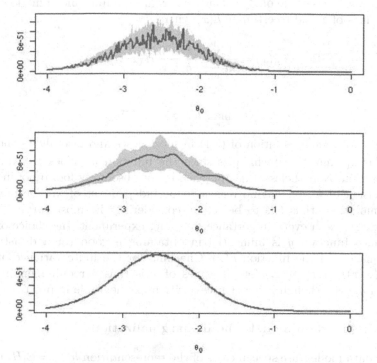

Fig. 5.11. Monte Carlo approximations of the marginal posterior of the probit posterior distribution associated with the `Pima.tr` dataset based on 10^3 simulations from a $t_5(\theta_1^m; \mu, \sigma)$ distribution. *(top)* Range of 100 replications of the approximation $\widehat{h}$ when simulating a different t sample for each value of θ_0 and overlay of one replication; *(middle)* range of 100 replications of the approximation $\widehat{h}$ when simulating the same t sample for each value of θ_0 and overlay of one replication; *(bottom)* comparison of the averages of both experiments *(the dotted graph corresponding to the top experiment is not distinguishable from the other graph).*

In order to handle vectors in the function `like`, we had to define it as

```
like=function(a,b){
  apply(pnorm(-a-outer(X=b,Y=da[,2],FUN="*"),lo=T)*(1-da[,1])
  +pnorm(a+outer(X=b,Y=da[,2],FUN="*"),lo=T)*da[,1],1,sum)}
```

due to the use of vectors in `pnorm`. Otherwise, it is impossible to apply functions like `integrate` to `exp(like(a,x))`. You can also check that the function `integrate` is completely unstable and thus untrustworthy in this example.

As illustrated by the example above, the type of approximation that is needed is a *uniformly convergent* approximation of h in order to trust $\widehat{h}(x)$ for any value of x. It thus makes sense instead to use importance sampling based on a *single* sample of Z_i's simulated from an importance function $g(z)$ for *all* values of x and to estimate $h(x)$ with

(5.8)
$$\hat{h}_m(x) = \frac{1}{m} \sum_{i=1}^{m} \frac{f(z_i|x)}{g(z_i)} H(x, z_i).$$

Solving
$$\max_x \hat{h}_m(x)$$

leads to a convergent solution of (5.1) in most cases and also allows the use of regular optimization techniques since the function $\hat{h}_m$ does not vary at each iteration. Nonetheless, this approach is not absolutely fool-proof in that the precision of $\hat{h}_m(x)$ has no reason to be independent of x. The number m of simulations thus has to be set by considering the most varying case. Furthermore, as in every importance sampling experiment, the choice of the importance function g is influential in obtaining a good (or a disastrous) approximation of the function $h(x)$. Checking for the finite variance of the ratio $f(z_i|x)H(x, z_i)/g(z_i)$ for all values of x is thus a requirement in the choice of g, even though it is not necessarily implementable in practice.

5.4.2 Missing-data models and demarginalization

Missing data models are special cases of the representation $h(x) = \mathbb{E}[H(x, Z)]$ best thought of as models where the density of the observations can be expressed as

(5.9)
$$g(x|\theta) = \int_{\mathcal{Z}} f(x, z|\theta) \, dz.$$

This representation occurs in many statistical settings, including censoring models and mixtures and latent variable models (tobit, probit, arch, stochastic volatility, etc.). Slice sampling (Section 7.4) is also an example of artificial missing data.

Example 5.12. The mixture model of Example 5.2 can be expressed as a missing-data model even though the (observed) likelihood can be computed in a manageable time. Indeed, if we introduce a vector $(z_1, \ldots, z_n) \in \{1, 2\}^n$ in addition to the sample $(x_1, \ldots, x_n)$ such that

$$P_\theta(Z_i = 1) = 1 - P_\theta(Z_i = 2) = 1/4, \quad X_i|Z_i = z \sim \mathcal{N}(\mu_z, 1),$$

we recover the mixture model (5.2) as the marginal distribution of X_i. The (observed) likelihood is then obtained as $\mathbb{E}[H(\mathbf{x}, \mathbf{Z})]$ for

$$H(\mathbf{x}, \mathbf{z}) \propto \prod_{i; \, z_i=1} \frac{1}{4} \exp\left\{-(x_i - \mu_1)^2/2\right\} \prod_{i; \, z_i=2} \frac{3}{4} \exp\left\{-(x_i - \mu_2)^2/2\right\},$$

where **x** denotes[2] $(x_1, \ldots, x_n)$ and **z** denotes $(z_1, \ldots, z_n) \in \{1, 2\}^n$. ◀

Example 5.13. Censored data may come from experiments where some potential observations are replaced with a lower bound because they take too long to observe. Suppose that we observe Y_1, ..., Y_m, iid, from $f(y - \theta)$ and that the $(n - m)$ remaining $(Y_{m+1}, \ldots, Y_n)$ are censored at the threshold a. The corresponding likelihood function is then

(5.10) $$L(\theta|\mathbf{y}) = [1 - F(a - \theta)]^{n-m} \prod_{i=1}^{m} f(y_i - \theta),$$

where F is the cdf associated with f and $\mathbf{y} = (y_1, \ldots, y_m)$. If we had observed the last $n - m$ values, say $\mathbf{z} = (z_{m+1}, \ldots, z_n)$, with $z_i \geq a$ $(i = m + 1, \ldots, n)$, we could have constructed the (complete data) likelihood

$$L^c(\theta|\mathbf{y}, \mathbf{z}) = \prod_{i=1}^{m} f(y_i - \theta) \prod_{i=m+1}^{n} f(z_i - \theta).$$

Note that

$$L(\theta|\mathbf{y}) = \mathbb{E}[L^c(\theta|\mathbf{y}, \mathbf{Z})] = \int_{\mathcal{Z}} L^c(\theta|\mathbf{y}, \mathbf{z}) f(\mathbf{z}|\mathbf{y}, \theta) \, d\mathbf{z},$$

where $f(\mathbf{z}|\mathbf{y}, \theta)$ is the density of the missing data conditional on the observed data, namely the product of the $f(z_i - \theta)/[1 - F(a - \theta)]$'s; i.e., $f(z - \theta)$ restricted to $(a, +\infty)$. ◀

From the perspective of this chapter, when (5.9) holds, the vector **Z** merely serves to simplify calculations, as it does not necessarily have a specific meaning for the corresponding statistical problem. It can still be seen as a *missing-data model* in the sense that **Z** can be interpreted as missing from the observations. We thus refer to the function $L^c(\theta|\mathbf{x}, \mathbf{z}) = f(\mathbf{x}, \mathbf{z}|\theta)$ as the "complete model" or "complete-data" likelihood, which is the likelihood we would obtain were we to observe $(\mathbf{x}, \mathbf{z})$, also called the *complete data*, albeit a misnomer since this is not truly data. This is a special case of *demarginalization*, a setting where a function of interest can be expressed as the integral of a more manageable quantity with no further constraint. We will meet such setups again in Chapter 7.

The following sections consider a hybrid strategy where the approximation of the objective function h and its maximization are grouped together in a single procedure. In the simplest cases, there is no randomness involved, and this leads to the EM algorithm, presented in Section 5.4.3. More general versions involving Monte Carlo steps are described in Section 5.4.4.

---

[2] In this section, in order to keep up with the traditional representation used for missing-data problems and the EM algorithm, we depart from our convention of denoting scalars and vectors with the same notation, using instead boldfaced symbols to represent samples of observed or latent variables.

5.4.3 The EM algorithm

The EM (which stands for expectation–maximization) algorithm is a deterministic optimization technique (Dempster et al., 1977) that takes advantage of the representation (5.9) to build a sequence of easier maximization problems whose limit is the answer to the original problem.

We thus assume that we observe $X_1, \ldots, X_n$, jointly distributed from $g(\mathbf{x}|\theta)$ that satisfies

$$g(\mathbf{x}|\theta) = \int_{\mathcal{Z}} f(\mathbf{x}, \mathbf{z}|\theta) \, d\mathbf{z},$$

and that we want to compute $\hat{\theta} = \arg\max L(\theta|\mathbf{x}) = \arg\max g(\mathbf{x}|\theta)$. Since the augmented data is $\mathbf{z}$, where $(\mathbf{X}, \mathbf{Z}) \sim f(\mathbf{x}, \mathbf{z}|\theta)$ the conditional distribution of the missing data $\mathbf{Z}$ given the observed data $\mathbf{x}$ is

$$k(\mathbf{z}|\theta, \mathbf{x}) = f(\mathbf{x}, \mathbf{z}|\theta)/g(\mathbf{x}|\theta).$$

Taking the logarithm of this expression leads to the following relationship between the complete-data likelihood $L^c(\theta|\mathbf{x}, \mathbf{z})$ and the observed-data likelihood $L(\theta|\mathbf{x})$. For any value θ_0,

$$(5.11) \qquad \log L(\theta|\mathbf{x}) = \mathbb{E}_{\theta_0}[\log L^c(\theta|\mathbf{x}, \mathbf{Z})] - \mathbb{E}_{\theta_0}[\log k(\mathbf{Z}|\theta, \mathbf{x})],$$

where the expectation is with respect to $k(\mathbf{z}|\theta_0, \mathbf{x})$. In the EM algorithm, while we aim at maximizing $\log L(\theta|\mathbf{x})$, only the first term on the right side of (5.11) will be considered.

Denoting

$$Q(\theta|\theta_0, \mathbf{x}) = \mathbb{E}_{\theta_0}[\log L^c(\theta|\mathbf{x}, \mathbf{Z})],$$

the EM algorithm indeed proceeds iteratively by maximizing $Q(\theta|\theta_0, \mathbf{x})$ at each iteration and, if $\hat{\theta}_{(1)}$ is the value of θ maximizing $Q(\theta|\theta_0, \mathbf{x})$, by replacing θ_0 by the updated value $\hat{\theta}_{(1)}$. In this manner, a sequence of estimators $\{\hat{\theta}_{(j)}\}_j$ is obtained, where $\hat{\theta}_{(j)}$ is defined as the value of θ maximizing $Q(\theta|\hat{\theta}_{(j-1)}, \mathbf{x})$; that is,

$$(5.12) \qquad Q(\hat{\theta}_{(j)}|\hat{\theta}_{(j-1)}, \mathbf{x}) = \max_{\theta} Q(\theta|\hat{\theta}_{(j-1)}, \mathbf{x}).$$

This iterative scheme thus contains both an expectation step and a maximization step, giving the algorithm its name.

Algorithm 3 The EM Algorithm
```
Pick a starting value θ̂(0)
Repeat
```

1. Compute (*the E-step*)

$$Q(\theta|\hat{\theta}_{(m)}, \mathbf{x}) = \mathbb{E}_{\hat{\theta}_{(m)}}[\log L^c(\theta|\mathbf{x}, \mathbf{Z})],$$

where the expectation is with respect to $k(\mathbf{z}|\hat{\theta}_{(m)}, \mathbf{x})$ and set $m = 0$.

2. Maximize $Q(\theta|\hat{\theta}_{(m)}, \mathbf{x})$ in θ and take (*the M-step*)

$$\hat{\theta}_{(m+1)} = \arg\max_{\theta} \ Q(\theta|\hat{\theta}_{(m)}, \mathbf{x})$$

and set $m = m + 1$

until a fixed point is reached; i.e., $\hat{\theta}_{(m+1)} = \hat{\theta}_{(m)}$.

By virtue of Jensen's inequality, it is easy (see, e.g., Robert and Casella, 2004, Theorem 5.15) to show that, at each step of the EM algorithm, the likelihood on the left side of (5.11) increases,

$$L(\hat{\theta}_{(j+1)}|\mathbf{x}) \geq L(\hat{\theta}_{(j)}|\mathbf{x}),$$

with equality holding if and only if $Q(\hat{\theta}_{(j+1)}|\hat{\theta}_{(j)}, \mathbf{x}) = Q(\hat{\theta}_{(j)}|\hat{\theta}_{(j)}, \mathbf{x})$. This means that, under some conditions, every limit point of an EM sequence $\{\hat{\theta}_{(j)}\}$ is a stationary point of $L(\theta|\mathbf{x})$, albeit not necessarily the maximum likelihood estimator or even a local maximum (see Wu, 1983, Theorem 3, or Boyles, 1983, for precise convergence results). It thus means that, in practice, running the EM algorithm several times with different, randomly chosen starting points is recommended if one wants to avoid using a poor approximation to the true maximum. (This is the only element of randomness involved in the EM algorithm since using the same starting point $\hat{\theta}_{(0)}$ ends up in the same fixed point.)

Implementing the EM algorithm thus means being able (a) to compute the function $Q(\theta'|\theta, \mathbf{x})$ and (b) to maximize this function. Numerous missing-data models can actually be processed that way, as illustrated in the following examples.

Example 5.14. (Continuation of Example 5.13) If the distribution $f(x - \theta)$ corresponds to the $\mathcal{N}(\theta, 1)$ distribution, the complete-data likelihood is

$$L^c(\theta|\mathbf{y}, \mathbf{z}) \propto \prod_{i=1}^{m} \exp\{-(y_i - \theta)^2/2\} \prod_{i=m+1}^{n} \exp\{-(z_i - \theta)^2/2\},$$

resulting in the expected complete-data log-likelihood

$$Q(\theta|\theta_0, \mathbf{y}) = -\frac{1}{2}\sum_{i=1}^{m}(y_i - \theta)^2 - \frac{1}{2}\sum_{i=m+1}^{n} \mathbb{E}_{\theta_0}[(Z_i - \theta)^2],$$

where the missing observations Z_i are distributed from a normal $\mathcal{N}(\theta, 1)$ distribution truncated in a. Doing the M-step (i.e., differentiating the function $Q(\theta|\theta_0, \mathbf{y})$ in θ) and setting it equal to 0 then leads to the EM update

$$\hat{\theta} = \frac{m\bar{y} + (n - m)\mathbb{E}_{\theta'}[Z_1]}{n}.$$

Since $\mathbb{E}_\theta[Z_1] = \theta + \frac{\varphi(a-\theta)}{1-\Phi(a-\theta)}$, where φ and Φ are the normal pdf and cdf, respectively, the EM sequence is

$$(5.13) \qquad \hat{\theta}^{(j+1)} = \frac{m}{n}\bar{y} + \frac{n - m}{n}\left[\hat{\theta}^{(j)} + \frac{\varphi(a - \hat{\theta}^{(j)})}{1 - \Phi(a - \hat{\theta}^{(j)})}\right].$$

The corresponding R implementation of the update is an easy recursion:

```
> theta=rnorm(1,mean=ybar,sd=sd(y))
> iteronstop=1
> while (nonstop){
+    theta=c(theta,m*ybar/n+(n-m)*(theta[iter]+
+    dnorm(a-theta[iter])/pnorm(a-theta[iter]))/n)
+    iter=iter+1
+    nonstop=(diff(theta[iter:(iter+1)])>10^(-4)) }
```

As can be checked when testing this program with arbitrary values of the parameters n, m, a, and $\bar{y}$, convergence to the maximum is quite rapid. Figure 5.12 illustrates this point by representing some sequences $\{\theta^{(j)}\}_j$ climbing the log-likelihood function. ◀

The example above is completely formal in that the observed likelihood can be computed in exact form, as shown by the graph in Figure 5.12, and optimized numerically. Thus, EM is not needed in this setting!

Exercise 5.8 Write an R code producing the likelihood (5.10) for a normal sample. Derive the numerical maximum likelihood estimator using optimise. Design a Monte Carlo experiment that studies the variation of the EM solutions around this numerical optimum.

Example 5.15. (Continuation of Example 5.12) Using the normal mixture likelihood of Examples 5.2 and 5.9, we saw in Figures 5.2 and 5.9 that the likelihood is bimodal when associated with a sample from the normal mixture (see Exercise 5.13 for a setup producing an alternative number of modes). Using the missing data structure exhibited in Example 5.12 leads to an objective function equal to

$$Q(\theta'|\theta, \mathbf{x}) = -\frac{1}{2}\sum_{i=1}^n \mathbb{E}_\theta\left[Z_i(x_i - \mu_1)^2 + (1 - Z_i)(x_i - \mu_2)^2 \,\middle|\, \mathbf{x}\right].$$

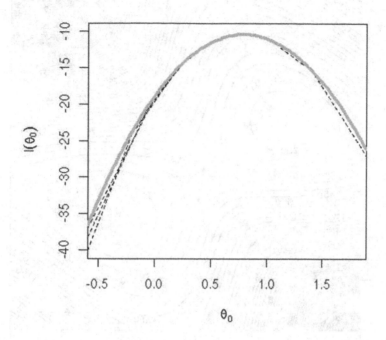

Fig. 5.12. Representation by dotted lines of 11 EM sequences started at random for the censored likelihood (5.10) when the data is normal, $\bar{y} = 0$, $a = 1$, $n = 30$, and $m = 20$, on top of the true log-likelihood *(grey curve)*.

Solving the M-step then provides the closed-form expressions

$$\mu_1' = \mathbb{E}_\theta \left[\sum_{i=1}^n Z_i x_i | \mathbf{x} \right] \Big/ \mathbb{E}_\theta \left[\sum_{i=1}^n Z_i | \mathbf{x} \right]$$

and

$$\mu_1' = \mathbb{E}_\theta \left[\sum_{i=1}^n (1 - Z_i) x_i | \mathbf{x} \right] \Big/ \mathbb{E}_\theta \left[\sum_{i=1}^n (1 - Z_i) | \mathbf{x} \right].$$

Since

$$\mathbb{E}_\theta \left[Z_i | \mathbf{x} \right] = \frac{\varphi(x_i - \mu_1)}{\varphi(x_i - \mu_1) + 3\varphi(x_i - \mu_2)},$$

the EM algorithm can easily be implemented in this setting. Running EM five times with various starting points chosen as in Figures 5.2 and 5.9, we represent in Figure 5.13 the corresponding occurrences. Two out of five sequences are attracted by the higher mode, while two others go to the lower mode (even

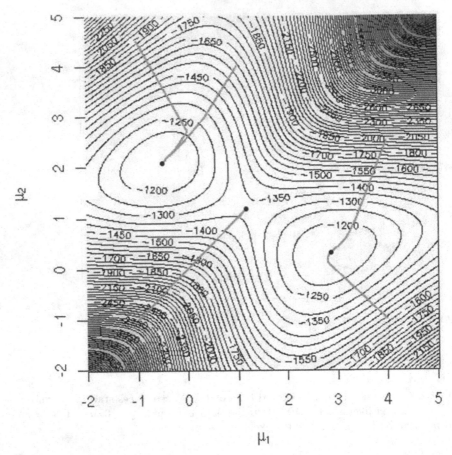

Fig. 5.13. Trajectories of five runs of the EM algorithm for Example 5.15 with their final position on the log-likelihood surface.

though the likelihood is considerably smaller) and, rather exceptionally (meaning that repeating the experiment on your own is unlikely to make this happen!), the fifth sequence ends up at the saddlepoint of the likelihood. Note that in a very few iterations the value is close to the modal value, with improvement brought by further iterations being incremental. For the current dataset, the EM sequences can thus only end up in three different spots, represented by dots on the likelihood surface. Note that, in comparison with the Newton–Raphson sequences in Figure 5.2, the convergence is systematically faster in that case.   ◀

Exercise 5.9 Show that, in Example 5.15, the weights $1/4$ and $3/4$ are irrelevant for the maximization of $Q(\theta'|\theta, \mathbf{x})$.

Exercise 5.10 Reproduce the analysis of Example 5.15 when the target is the normal scale mixture (5.2) by building the complete likelihood based on the allocation vector $\mathbf{z} = (z_1, \ldots, z_n)$ and by deriving the updating step of the corresponding EM algorithm. Implement the algorithm on the dataset `log(deaths)` (available in `library(MASS)`).

This example reinforces the need for rerunning the algorithm a number of times, each time starting from a different initial value that can be chosen randomly, provided the range of the starting distribution is wide enough to encompass all possible modes.

5.4.4 Monte Carlo EM

A difficulty with the implementation of the EM algorithm is that each E-step requires the computation of the expected log-likelihood $Q(\theta|\theta_0, \mathbf{x})$. Besides standard cases where the function Q is available in closed form, since Q is naturally expressed as an expectation, it can be approximated using Monte Carlo following the general rules of Section 5.4.1. This means for instance simulating $\mathbf{Z}_1, \ldots, \mathbf{Z}_T$ from the conditional distribution $k(\mathbf{z}|\mathbf{x}, \hat{\theta}_{(m)})$ and then maximizing the approximate complete-data log-likelihood

$$(5.14) \qquad \hat{Q}(\theta|\theta_0, \mathbf{x}) = \frac{1}{T} \sum_{i=1}^{T} \log L^c(\theta|\mathbf{x}, \mathbf{z}_i)$$

as suggested by Wei and Tanner (1990) under the name of Monte Carlo EM (MCEM). However, as described in Section 5.4.1, a more stable version relies on importance sampling in order to avoid simulating a new sample for each new value $\hat{\theta}_{(m)}$. A specific solution (Geyer and Thompson, 1992) consists in using only the conditional distribution $k(\mathbf{z}|\mathbf{x}, \hat{\theta}_{(0)})$ due to the identity

$$\arg \max_{\theta} L(\theta|\mathbf{x}) = \arg \max_{\theta} \log \frac{g(\mathbf{x}|\theta)}{g(\mathbf{x}|\theta_{(0)})} = \arg \max_{\theta} \log \mathbb{E}_{\theta_{(0)}} \left[\left. \frac{f(\mathbf{x}, \mathbf{z}|\theta)}{f(\mathbf{x}, \mathbf{z}|\theta_{(0)})} \right| \mathbf{x} \right],$$

which implies that simulating $\mathbf{Z}_1, \ldots, \mathbf{Z}_T$ from the conditional distribution $k(\mathbf{z}|\mathbf{x}, \hat{\theta}_{(0)})$ provides as an approximation to the log-likelihood (modulo an additive constant)

$$\log L(\theta|\mathbf{x}) \approx \frac{1}{T} \sum_{i=1}^{T} \frac{L^c(\theta|\mathbf{x}, \mathbf{z}_i)}{L^c(\theta_{(0)}|\mathbf{x}, \mathbf{z}_i)},$$

the stability of which depends on the choice of $\theta_{(0)}$ in order to ensure a finite variance for this approximation.

Example 5.16. A classic example of the EM algorithm is a genetics problem (Dempster et al., 1977) where observations (x_1, x_2, x_3, x_4) are gathered from the multinomial distribution

$$\mathcal{M}\left(n; \frac{1}{2} + \frac{\theta}{4}, \frac{1}{4}(1 - \theta), \frac{1}{4}(1 - \theta), \frac{\theta}{4}\right).$$

Estimation is easier if the x_1 cell is split into two cells, so we create the augmented model

$$(z_1, z_2, x_2, x_3, x_4) \sim \mathcal{M}\left(n; \frac{1}{2}, \frac{\theta}{4}, \frac{1}{4}(1 - \theta), \frac{1}{4}(1 - \theta), \frac{\theta}{4}\right)$$

with $x_1 = z_1 + z_2$. The complete-data likelihood function is then simply $\theta^{z_2 + x_4}(1 - \theta)^{x_2 + x_3}$, as opposed to the observed-data likelihood function $(2 + \theta)^{x_1}\theta^{x_4}(1 - \theta)^{x_2 + x_3}$. The expected complete log-likelihood function is

$$\mathbb{E}_{\theta_0}[(Z_2 + x_4) \log \theta + (x_2 + x_3) \log(1 - \theta)]$$
$$= \left(\frac{\theta_0}{2 + \theta_0} x_1 + x_4\right) \log \theta + (x_2 + x_3) \log(1 - \theta),$$

which can easily be maximized in θ, leading to the EM step

$$\hat{\theta}_1 = \left\{\frac{\theta_0 x_1}{2 + \theta_0 + x_4}\right\} \Big/ \left\{\frac{\theta_0 x_1}{2 + \theta_0} + x_2 + x_3 + x_4\right\}.$$

A Monte Carlo EM solution would replace the expectation $\theta_0 x_1/(2 + \theta_0)$ with the empirical average

$$\bar{z}_m = \frac{1}{m} \sum_{i=1}^{m} z_i,$$

where the z_i's are simulated from a binomial distribution $\mathcal{B}(x_1, \theta_0/(2 + \theta_0))$ or, equivalently, by

$$m\bar{z}_m \sim \mathcal{B}(mx_1, \theta_0/(2 + \theta_0)).$$

The MCEM step would then be

$$\hat{\hat{\theta}}_1 = \frac{\bar{z}_m + x_4}{\bar{z}_m + x_2 + x_3 + x_4},$$

which obviously converges to $\hat{\theta}_1$ as m grows to infinity.   ◀

Exercise 5.11 In the setting of Example 5.16, starting from $\theta_0 = .5$, design a Monte Carlo experiment to evaluate the variation of the MCEM sequence around the EM sequence for $m = 100$ and plot the range of variation using `polygon`.

This example is merely a formal illustration of the Monte Carlo EM algorithm and its convergence properties since EM can be applied. (In Example

7.8, we will revisit this example with a Gibbs sampler.) The next example, however, details a situation in which the E-step is too complicated to be implemented and where the Monte Carlo EM algorithm provides a realistic (if not straightforward) alternative.

Example 5.17. A simple random effect logit model processed in Booth and Hobert (1999) represents observations y_{ij} $(i = 1, \ldots, n, j = 1, \ldots, m)$ as distributed conditionally on one covariate x_{ij} as a logit model

$$P(y_{ij} = 1 | x_{ij}, u_i, \beta) = \frac{\exp\{\beta x_{ij} + u_i\}}{1 + \exp\{\beta x_{ij} + u_i\}},$$

where $u_i \sim \mathcal{N}(0, \sigma^2)$ is an unobserved random effect. The vector of random effects $(U_1, \ldots, U_n)$ therefore corresponds to the missing data $\mathbf{Z}$. When considering the function $Q(\theta' | \theta, \mathbf{x}, \mathbf{y})$,

$$Q(\theta' | \theta, \mathbf{x}, \mathbf{y}) = \sum_{i,j} y_{ij} \mathbb{E}[\beta' x_{ij} + U_i | \beta, \sigma, \mathbf{x}, \mathbf{y}]$$

$$- \sum_{i,j} \mathbb{E}[\log 1 + \exp\{\beta' x_{ij} + U_i\} | \beta, \sigma, \mathbf{x}, \mathbf{y}]$$

$$- \sum_i \mathbb{E}[U_i^2 | \beta, \sigma, \mathbf{x}, \mathbf{y}] / 2\sigma'^2 - n \log \sigma',$$

with $\theta = (\beta, \sigma)$, it is impossible to compute the expectations in U_i. Were those available, the M-step would then be almost straightforward since maximizing $Q(\theta' | \theta, \mathbf{x}, \mathbf{y})$ in σ' leads to

$$\sigma'^2 = \frac{1}{n} \sum_i \mathbb{E}[U_i^2 | \beta, \sigma, \mathbf{x}, \mathbf{y}],$$

while maximizing $Q(\theta' | \theta, \mathbf{x}, \mathbf{y})$ in β' produces the fixed-point equation

$$\sum_{i,j} y_{ij} x_{ij} = \sum_{i,j} \mathbb{E}\left[\frac{\exp\{\beta' x_{ij} + U_i\}}{1 + \exp\{\beta' x_{ij} + U_i\}} \middle| \beta, \sigma, \mathbf{x}, \mathbf{y}\right] x_{ij},$$

which is not particularly easy to solve in β.

   The alternative to EM is therefore to simulate the U_i's conditional on $\beta, \sigma, \mathbf{x}, \mathbf{y}$ in order to replace the expectations above with Monte Carlo approximations. While a direct simulation from

(5.15) $$\pi(u_i | \beta, \sigma, \mathbf{x}, \mathbf{y}) \propto \frac{\exp\left\{\sum_j y_{ij} u_i - u_i^2 / 2\sigma^2\right\}}{\prod_j [1 + \exp\{\beta x_{ij} + u_i\}]}$$

is feasible (Booth and Hobert, 1999), it requires some preliminary tuning better avoided at this stage, and it is thus easier to implement an MCMC version of the simulation of the u_i's toward the approximations of both expectations. We will discuss how to construct a (standard) MCMC algorithm to produce samples of u_i's in the next chapter and we assume from now on that, at each iteration of the MCEM algorithm, we obtain an (n, T) matrix

```
> mcmc(beta,sigma,x,y,T)
```

whose ith row is a sample of u_{it}'s distributed from (5.15). (Both x and y are defined as (n, m) matrices.) Given this matrix of samples, we can update σ^2 by sum(u^2)/(n*T) and, for updating β, we can for instance use the function uniroot for the target function

```
targ=function(beta,x,y,uni){
    xs=exp(beta*x)
    xxs=x*xs
    ome=exp(uni)
    prodct=0
    for (j in 1:m) for (t in 1:T)
        prodct=prodct+sum(xxs[,j]*ome[,t]/(1+xs[,j]*ome[,t]))
    prodct-sum(T*x*y)
    }
```

as in

```
> beta=uniroot(targ,x=x,y=y,u=u,int=mlan+10*sigma*c(-1,1))
```

if mlan is defined as the MLE of β when there is no random effect:

```
> mlan=as.numeric(glm(as.vector(y)~as.vector(x)-1,
+       fa=binomial)$coe)
```

The entire MCEM step is therefore produced by

```
> T=1000        #Number of MCEM simulations
> beta=mlan
> sigma=diff=iter=factor=1
> while (diff>10^-3){
+    samplu=mcmc(beta[iter],sigma[iter],x,y,T)
+    sigma=c(sigma,sd(as.vector(samplu)))
+    beta=c(beta,uniroot(targ,x=x,y=y,u=samplu,
+          inter=mlan+c(-10*sigma,10*sigma)))
+    diff=max(abs(diff(beta[iter:(iter+1)])),
+          abs(diff(sigma[iter:(iter+1)])))
+    iter=iter+1
+    T=T*2}
```

where the last line T=T*2 of the while loop is intended to stabilize the MCEM sequence by increasing the number of Monte Carlo steps at each iteration, as suggested in Wei and Tanner (1990), McCulloch (1997), and Booth and Hobert (1999). The stopping rule is based on this stabilization of the MCEM values, even though an alternative criterion based on the Monte Carlo approximation to the (observed) likelihood

```
like=function(beta,sigma){
  lo=0
  for (t in 1:(10*T)){
    uu=rnorm(n)*sigma
    lo=lo+exp(sum(as.vector(y)*(beta*as.vector(x)+rep(uu,m)))-
          sum(log(1+exp(beta*as.vector(x)+rep(uu,m))))) }
  lo/T
  }
```

could be used as well. Figure 5.14 shows the sequence of θ's produced by this algorithm as well as how the sequence of completed likelihoods evaluated at the true random effects evolves with β.   ◄

In the example above, when using the R procedure uniroot in conjunction with the target function targ, the last argument of targ cannot be chosen to be u, as in

```
> targ=function(beta,x,y,u)
```

because it would then create confusion with the upper argument of uniroot

```
> uniroot(targ,int=c(-5*sigma0,5*sigma0),x=x,y=y,u=samplu)
Error in f(lower, ...) : argument "u" is missing, with
no default
```

while using the argument uni avoids the confusion:

```
> uniroot(targ,int=c(-5*sigma0,5*sigma0),x=x,y=y,uni=samplu)
$root
[1] -2.995976
$f.root
[1] 0.003151796
$iter
[1] 10
$estim.prec
[1] 6.103516e-05
```

Exercise 5.12 Show that (5.15) can be simulated using an Accept–Reject algorithm based on a normal proposal. Examine the performance of this algorithm in terms of acceptance probability when using a simulated sample with the same parameters as in Figure 5.14. (*Hint:* See Booth and Hobert (1999) for a version that does not require a different maximization for each new value of β, as well as an importance sampling alternative.)

↯ Note that the MCEM algorithm no longer enjoys the fundamental EM monotonicity property. It is therefore important to assess that the se-

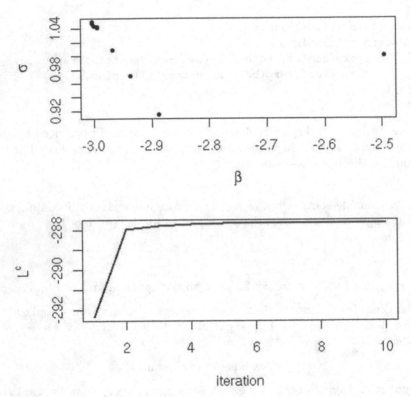

Fig. 5.14. MCEM estimation of the parameters of a logit model with random effects based on a simulated dataset with $n = 20$, $m = 35$, $\beta = -3$, $\sigma = 1$, and x_{ij} randomly distributed in $\{-1, 0, 1\}$. The top graph represents the sequence of (β, σ)'s produced by the MCEM algorithm when starting from $\beta_0 = \hat{\beta}$, the MLE estimator for the logit model, and $\sigma_0 = 1$. The final value for the approximation of the MLE is $(\beta_{10}, \sigma_{10}) = (-3.002, 1.048)$. The bottom graph provides the sequence of $L^c(\beta, \sigma, \mathbf{u}|\mathbf{x}, \mathbf{y})$ for the simulated vector $\mathbf{u}$ of random effects.

quence of (β, σ)'s produced by the MCEM algorithm converges to an approximate maximum of the model likelihood, either by numerically evaluating the likelihood or by repeating the runs with different starting values. In contrast with more generic Monte Carlo methods, the MCEM algorithm nonetheless offers the advantage of approximating the converging EM sequence, rather than maximizing an approximation of the likelihood.

In contrast with EM, as MCEM is a Monte Carlo method, its outcome should be associated with an error evaluation. Besides using the crude and costly device of path replications, Booth and Hobert (1999) provide a first-order approximation based on a Taylor expansion of Q:

$$\text{var}(\theta_1|\theta_0,\mathbf{x}) \approx \left[\frac{\partial^2 Q(\theta_1|\theta_0,\mathbf{x})}{\partial\theta\partial\theta^{\mathrm{T}}}\right]^{-1} \text{var}\left(\frac{\partial Q(\theta_1|\theta_0,\mathbf{x})}{\partial\theta}\right)\left[\frac{\partial^2 Q(\theta_1|\theta_0,\mathbf{x})}{\partial\theta\partial\theta^{\mathrm{T}}}\right]^{-1}.$$

The inner variance term can then be easily evaluated based on the simulated missing variables.

5.5 Additional exercises

Exercise 5.13 In the setting of Example 5.2, draw the likelihood surface associated with the model (5.2) when the sample of size 400 is produced by

```
> x=rnorm(80,mean=-4)
> for (i in 1:4) x=c(x,rnorm(80,mean=-4+2*i))
```

and determine the number of local maxima associated with a given sample.

Exercise 5.14 Consider a sample of size n from a mixture distribution with unknown weights,
$$X_i \sim \theta g(x) + (1-\theta)h(x), \quad i = 1,\ldots,n,$$
where $g(\cdot)$ and $h(\cdot)$ are known.

a. Introduce $Z_1,\ldots,Z_n$, where Z_i indicates the distribution from which X_i has been drawn, so
$$X_i|Z_i = 1 \sim g(x), \quad X_i|Z_i = 0 \sim h(x).$$
Show that the complete-data likelihood can be written as
$$L^c(\theta|\mathbf{x},\mathbf{z}) = \prod_{i=1}^{n} [z_i g(x_i) + (1-z_i)h(x_i)]\, \theta^{z_i}(1-\theta)^{1-z_i}.$$

b. Show that $\mathbb{E}[Z_i|\theta,x_i] = \theta g(x_i)/[\theta g(x_i) + (1-\theta)h(x_i)]$, and deduce that the EM sequence is given by
$$\hat{\theta}_{(j+1)} = \frac{1}{n}\sum_{i=1}^{n}\frac{\hat{\theta}_{(j)}g(x_i)}{\hat{\theta}_{(j)}g(x_i) + (1-\hat{\theta}_{(j)})h(x_i)}.$$

c. Examine the convergence properties of this EM algorithm on a simulated dataset with $n = 25$, $\theta = 0.3$, $h(x) = \varphi(x)$, and $g(x) = \varphi((x-2)/2)/2$, where φ denotes the $\mathcal{N}(0,1)$ density.

Exercise 5.15 Consider the sample $\mathbf{x} = (0.12, 0.17, 0.32, 0.56, 0.98, 1.03, 1.10, 1.18, 1.23, 1.67, 1.68, 2.33)$, generated from an exponential mixture
$$p\,\mathcal{E}xp(\lambda) + (1-p)\,\mathcal{E}xp(\mu).$$

All parameters p, μ, λ are unknown.

a. Show that the likelihood $h(p, \lambda, \mu)$ can be expressed as $\mathbb{E}[H(x, Z)]$, where $z = (z_1, \ldots, z_{12})$ corresponds to the vector of allocations of the observations x_i to the first and second components of the mixture; that is, for $i = 1, \ldots, 12$,

$$P(z_i = 1) = 1 - P(z_i = 2) = \frac{p\lambda \exp(-\lambda x_i)}{p\lambda \exp(-\lambda x_i) + (1-p)\mu \exp(-\mu x_i)}.$$

b. Construct an EM algorithm for this model, and derive the maximum likelihood estimators of the parameters for the sample provided above.

Exercise 5.16 Consider the function

$$h(\theta) = \frac{||\theta||^2 (p + ||\theta||^2)(2p - 2 + ||\theta||^2)}{(1 + ||\theta||^2)(p + 1 + ||\theta||^2)(p + 3 + ||\theta||^2)}$$

when $\theta \in \mathbb{R}^p$ and $p = 10$.

a. Show that the function $h(\theta)$ has a unique maximum in $||\theta||^2$.
b. Show that $h(\theta)$ can be expressed as $\mathbb{E}[H(\theta, Z)]$, where $z = (z_1, z_2, z_3)$ and $Z_i \sim \mathcal{E}xp(1/2)$ $(i = 1, 2, 3)$. Deduce that $f(z|x)$ does not depend on x in (5.8).
c. When $g(z) = \exp(-\alpha\{z_1 + z_2 + z_3\})$, show that the variance of (5.8) is infinite for some values of $t = ||\theta||^2$ when $\alpha > 1/2$. Identify A_2, the set of values of t for which the variance of (5.8) is infinite when $\alpha = 2$.
d. Study the behavior of the estimate (5.8) when t goes from A_2 to its complement A_2^c to see if the infinite variance can be detected in the evaluation of $h(t)$.

Exercise 5.17 Referring to Example 5.14,

a. Give the density of the missing data $(Z_{n-m+1}, \ldots, Z_n)$.
b. Show that $\mathbb{E}_{\theta'}[Z_i] = \theta' + \frac{\phi(a - \theta')}{1 - \Phi(a - \theta')}$ and that the EM sequence is given by (5.13).

Exercise 5.18 In the setup of the probit model of Example 5.11, denoting $\beta = (\theta_0, \theta_1)$ and replacing the covariate x with the vector $x = (x, 1)$,

a. Give the likelihood associated with a sample $((x_1, y_1), \ldots, (x_n, y_n))$.
b. Show that, if we associate with each observation (x_i, y_i) a missing variable Z_i such that

$$Z_i | X_i = x \sim \mathcal{N}(x^T \beta, 1) \qquad Y_i = \mathbb{I}_{Z_i > 0},$$

iteration m of the associated EM algorithm is the expected least squares estimator

$$\beta_{(m)} = (X^T X)^{-1} X^T \mathbb{E}_{\beta_{(m-1)}}[\mathbf{Z}|\mathbf{x}, \mathbf{y}],$$

where $\mathbf{x} = (x_1, \ldots, x_n)$, $\mathbf{y} = (y_1, \ldots, y_n)$ and $\mathbf{Z} = (Z_1, \ldots, Z_n)^T$, and X is the matrix with columns made of the x_i's.
c. Give the value of $\mathbb{E}_{\beta}[Z_i | x_i, y_i]$.
d. Implement this EM algorithm for the `Pima.tr` dataset when x corresponds to the variables `glu`, `skin`, and `bmi`, respectively. Compare it with the solutions provided by `glm`.

Exercise 5.19 Test the limitations of the `uniroot` function when considering $h(x) = (x - 3)(x + 6)(1 + \sin(60x))$ on the intervals $(-2, 10)$ and $(-8, -1)$.

Exercise 5.20 An alternate implementation of the Monte Carlo EM might be, for $Z_1, \ldots, Z_m \sim k(\mathbf{z}|\mathbf{x}, \theta)$, to iteratively maximize

$$\log \hat{L}(\theta|\mathbf{x}) = \frac{1}{m} \sum_{i=1}^{m} \{\log L^c(\theta|\mathbf{x}, \mathbf{z}_i) - \log k(\mathbf{z}_i|\theta, x)\}$$

(which might more accurately be called Monte Carlo maximum likelihood).

a. Show that $\hat{L}(\theta|\mathbf{x}) \to L(\theta|\mathbf{x})$ as $m \to \infty$.
b. Show how to use $\hat{L}(\theta|\mathbf{x})$ to obtain the MLE in Example 5.15 and illustrate the convergence of this procedure for the data faithful[,1].

Exercise 5.21 For the situation in Example 5.16, data $(x_1, x_2, x_3, x_4) = (125, 18, 20, 34)$ are collected.

a. Construct an EM algorithm to find the MLE of θ.
b. Construct a Monte Carlo EM algorithm to find the MLE of θ. Compare your results with those of part a using a Monte Carlo experiment evaluating the variability of the MCEM sequence.

Exercise 5.22 The following dataset gives independent observations of $Z = (X, Y) \sim \mathcal{N}_2(0, \Sigma)$ with missing data denoted by $*$.

x	1.17	-0.98	0.18	0.57	0.21	*	*	*
y	0.34	-1.24	-0.13	*	*	-0.12	-0.83	1.64

a. Show that the observed likelihood is

$$\prod_{i=1}^{3} \left\{ |\Sigma|^{-1/2} e^{-z_i^t \Sigma^{-1} z_i/2} \right\} \sigma_1^{-2} e^{-(x_4^2+x_5^2)/2\sigma_1^2} \sigma_2^{-3} e^{-(y_6^2+y_7^2+y_8^2)/2\sigma_2^2}.$$

b. Examine the consequence of the choice of $\pi(\Sigma) \propto |\Sigma|^{-1}$ on the posterior distribution of Σ.
c. Show that the missing data can be simulated from

$$X_i^* \sim \mathcal{N}\left(\rho \frac{\sigma_1}{\sigma_2} y_i, \sigma_1^2(1 - \rho^2)\right) \qquad (i = 6, 7, 8),$$

$$Y_i^* \sim \mathcal{N}\left(\rho \frac{\sigma_2}{\sigma_1} x_i, \sigma_2^2(1 - \rho^2)\right) \qquad (i = 4, 5),$$

to derive a Monte Carlo EM algorithm.
d. Build an efficient simulation method to obtain the MLE of the covariance matrix Σ.

6

Metropolis–Hastings Algorithms

"How absurdly simple!", I cried.
"Quite so!", said he, a little nettled. "Every problem becomes very childish when once it is explained to you."

Arthur Conan Doyle
The Adventure of the Dancing Men

Reader's guide

This chapter is the first of a series of two on simulation methods based on *Markov chains*. Although the Metropolis–Hastings algorithm can be seen as one of the most general Markov chain Monte Carlo (MCMC) algorithms, it is also one of the simplest both to understand and explain, making it an ideal algorithm to start with.

This chapter begins with a quick refresher on Markov chains, just the basics needed to understand the algorithms. Then we define the Metropolis–Hastings algorithm, focusing on the most common versions of the algorithm. We end up discussing the calibration of the algorithm via its acceptance rate in Section 6.5.

C.P. Robert, G. Casella, *Introducing Monte Carlo Methods with R*, Use R,
DOI 10.1007/978-1-4419-1576-4_6, © Springer Science+Business Media, LLC 2010

6.1 Introduction

For reasons that will become clearer as we proceed, we now make a fundamental shift in the choice of our simulation strategy. Up to now we have typically generated *iid* variables directly from the density of interest f or indirectly in the case of importance sampling. The Metropolis–Hastings algorithm introduced below instead generates *correlated* variables from a Markov chain. The reason why we opt for such a radical change is that Markov chains carry different convergence properties that can be exploited to provide easier proposals in cases where generic importance sampling does not readily apply. For one thing, the requirements on the target f are quite minimal, which allows for settings where very little is known about f. Another reason, as illustrated in the next chapter, is that this Markov perspective leads to efficient decompositions of high-dimensional problems in a sequence of smaller problems that are much easier to solve.

Thus, be warned that this is a pivotal chapter in that we now introduce a totally new perspective on the generation of random variables, one that has had a profound effect on research and has expanded the application of statistical methods to solve more difficult and more relevant problems in the last twenty years, even though the origins of those techniques are tied with those of the Monte Carlo method in the remote research center of Los Alamos during the Second World War. Nonetheless, despite the recourse to Markov chain principles that are briefly detailed in the next section, the implementation of these new methods is not harder than those of earlier chapters, and there is no need to delve any further into Markov chain theory, as you will soon discover. (Most of your time and energy will be spent in designing and assessing your MCMC algorithms, just as for the earlier chapters, not in establishing convergence theorems, so take it easy!)

6.2 A peek at Markov chain theory

↯ This section is intended as a minimalist refresher on Markov chains in order to define the vocabulary of Markov chains, nothing more. In case you have doubts or want more details about these notions, you are strongly advised to check a more thorough treatment such as Robert and Casella (2004, Chapter 6) or Meyn and Tweedie (1993) since no theory of convergence is provided in the present book.

A Markov chain $\{X^{(t)}\}$ is a sequence of dependent random variables

$$X^{(0)}, X^{(1)}, X^{(2)}, \ldots, X^{(t)}, \ldots$$

such that the probability distribution of $X^{(t)}$ given the past variables depends only on $X^{(t-1)}$. This conditional probability distribution is called a *transition kernel* or a *Markov kernel K*; that is,

$$X^{(t+1)} \mid X^{(0)}, X^{(1)}, X^{(2)}, \ldots, X^{(t)} \sim K(X^{(t)}, X^{(t+1)}).$$

For example, a simple *random walk* Markov chain satisfies

$$X^{(t+1)} = X^{(t)} + \epsilon_t,$$

where $\epsilon_t \sim \mathcal{N}(0, 1)$, independently of $X^{(t)}$; therefore, the Markov kernel $K(X^{(t)}, X^{(t+1)})$ corresponds to a $\mathcal{N}(X^{(t)}, 1)$ density.

For the most part, the Markov chains encountered in Markov chain Monte Carlo (MCMC) settings enjoy a very strong stability property. Indeed, a *stationary probability distribution* exists by construction for those chains; that is, there exists a probability distribution f such that if $X^{(t)} \sim f$, then $X^{(t+1)} \sim f$. Therefore, formally, the kernel and stationary distribution satisfy the equation

$$(6.1) \qquad \int_{\mathcal{X}} K(x, y) f(x) \mathrm{d}x = f(y).$$

The existence of a stationary distribution (or *stationarity*) imposes a preliminary constraint on K called *irreducibility* in the theory of Markov chains, which is that the kernel K allows for free moves all over the stater-space, namely that, no matter the starting value $X^{(0)}$, the sequence $\{X^{(t)}\}$ has a positive probability of eventually reaching any region of the state-space. (A sufficient condition is that $K(x, \cdot) > 0$ everywhere.) The existence of a stationary distribution has major consequences on the behavior of the chain $\{X^{(t)}\}$, one of which being that most of the chains involved in MCMC algorithms are *recurrent*, that is, they will return to any arbitrary nonnegligible set an infinite number of times.

Exercise 6.1 Consider the Markov chain defined by $X^{(t+1)} = \varrho X^{(t)} + \epsilon_t$, where $\epsilon_t \sim \mathcal{N}(0, 1)$. Simulating $X^{(0)} \sim \mathcal{N}(0, 1)$, plot the histogram of a sample of $X^{(t)}$ for $t \leq 10^4$ and $\varrho = .9$. Check the potential fit of the stationary distribution $\mathcal{N}(0, 1/(1 - \varrho^2))$.

In the case of recurrent chains, the stationary distribution is also a *limiting distribution* in the sense that the limiting distribution of $X^{(t)}$ is f for almost any initial value $X^{(0)}$. This property is also called *ergodicity*, and it obviously has major consequences from a simulation point of view in that, if a given kernel K produces an ergodic Markov chain with stationary distribution f, generating a chain from this kernel K will eventually produce simulations from f. In particular, for integrable functions h, the standard average

$$(6.2) \qquad \frac{1}{T} \sum_{t=1}^{T} h(X^{(t)}) \longrightarrow \mathbb{E}_f[h(X)],$$

which means that the Law of Large Numbers that lies at the basis of Monte Carlo methods (Section 3.2) can also be applied in MCMC settings. (It is then sometimes called the *Ergodic Theorem*.)

We won't dabble any further into the theory of convergence of MCMC algorithms, relying instead on the guarantee that standard versions of these algorithms such as the Metropolis–Hastings algorithm or the Gibbs sampler are almost always theoretically convergent. Indeed, the real issue with MCMC algorithms is that, despite those convergence guarantees, the practical implementation of those principles may imply a very lengthy convergence time or, worse, may give an impression of convergence while missing some important aspects of f, as discussed in Chapter 8.

There is, however, one case where convergence never occurs, namely when, in a Bayesian setting, the posterior distribution is not proper (Robert, 2001) since the chain cannot be recurrent. With the use of improper priors $f(x)$ being quite common in complex models, there is a possibility that the product likelihood × prior, $\ell(x) \times f(x)$, is not integrable and that this problem goes undetected because of the inherent complexity. In such cases, Markov chains can be simulated in conjunction with the target $\ell(x) \times f(x)$ but cannot converge. In the best cases, the resulting Markov chains will quickly exhibit divergent behavior, which signals there is a problem. Unfortunately, in the worst cases, these Markov chains present all the outer signs of stability and thus fail to indicate the difficulty. More details about this issue are discussed in Section 7.6.4 of the next chapter.

Exercise 6.2 Show that the random walk has no stationary distribution. Give the distribution of $X^{(t)}$ for $t = 10^4$ and $t = 10^6$ when $X^{(0)} = 0$, and deduce that $X^{(t)}$ has no limiting distribution.

6.3 Basic Metropolis–Hastings algorithms

The working principle of Markov chain Monte Carlo methods is quite straightforward to describe. Given a target density f, we build a Markov kernel K with stationary distribution f and then generate a Markov chain $(X^{(t)})$ using this kernel so that the limiting distribution of $(X^{(t)})$ is f and integrals can be approximated according to the Ergodic Theorem (6.2). The difficulty should thus be in constructing a kernel K that is associated with an arbitrary density f. But, quite miraculously, there exist methods for deriving such kernels that are universal in that they are theoretically valid for any density f!

The Metropolis–Hastings algorithm is an example of those methods. (Gibbs sampling, described in Chapter 7, is another example with equally universal potential.) Given the target density f, it is associated with a working conditional density $q(y|x)$ that, in practice, is easy to simulate. In addition, q can be almost arbitrary in that the only theoretical requirements are that the ratio $f(y)/q(y|x)$ is known up to a constant *independent* of x and that $q(\cdot|x)$ has enough dispersion to lead to an exploration of the entire support of f. Once again, we stress the incredible feature of the Metropolis–Hastings algorithm that, for *every* given q, we can then construct a Metropolis–Hastings kernel such that f is its stationary distribution.

6.3.1 A generic Markov chain Monte Carlo algorithm

The Metropolis–Hastings algorithm associated with the objective (target) density f and the conditional density q produces a Markov chain $(X^{(t)})$ through the following transition kernel:

Algorithm 4 Metropolis–Hastings
Given $x^{(t)}$,
  1. Generate $Y_t \sim q(y|x^{(t)})$.
  2. Take

$$X^{(t+1)} = \begin{cases} Y_t & \text{with probability} \quad \rho(x^{(t)}, Y_t), \\ x^{(t)} & \text{with probability} \quad 1 - \rho(x^{(t)}, Y_t), \end{cases}$$

  where

$$\rho(x, y) = \min\left\{\frac{f(y)}{f(x)} \frac{q(x|y)}{q(y|x)}, 1\right\}.$$

A generic R implementation is straightforward, assuming a generator for $q(y|x)$ is available as geneq(x). If x[t] denotes the value of $X^{(t)}$,

```
> y=geneq(x[t])
> if (runif(1)<f(y)*q(y,x[t])/(f(x[t])*q(x[t],y))){
+     x[t+1]=y
+     }else{
+     x[t+1]=x[t]
+     }
```

since the value y is always accepted when the ratio is larger than one.

The distribution q is called the *instrumental* (or *proposal* or *candidate*) *distribution* and the probability $\rho(x, y)$ the *Metropolis–Hastings acceptance probability*. It is to be distinguished from the *acceptance rate*, which is the average of the acceptance probability over iterations,

$$\bar{\rho} = \lim_{T \to \infty} \frac{1}{T} \sum_{t=0}^{T} \rho(X^{(t)}, Y_t) = \int \rho(x, y) f(x) q(y|x) \, dy dx.$$

This quantity allows an evaluation of the performance of the algorithm, as discussed in Section 6.5.

While, at first glance, Algorithm 4 does not seem to differ from Algorithm 2, except for the notation, there are two fundamental differences between the two algorithms. The first difference is in their use since Algorithm 2 aims at maximizing a function $h(x)$, while the goal of Algorithm 4 is to explore the support of the density f according to its probability. The second difference is in their convergence properties. With the proper choice of a temperature schedule T_t in Algorithm 2, the simulated annealing algorithm converges to the maxima of the function h, while the Metropolis–Hastings algorithm is converging to the distribution f itself. Finally, modifying the proposal q along iterations may have drastic consequences on the convergence pattern of this algorithm, as discussed in Section 8.5.

Algorithm 4 satisfies the so-called *detailed balance condition*,

$$(6.3) \qquad\qquad f(x)K(y|x) = f(y)K(x|y)\,,$$

from which we can deduce that f is the stationary distribution of the chain $\{X^{(t)}\}$ by integrating each side of the equality in x (see Exercise 6.8).

That Algorithm 4 is naturally associated with f as its stationary distribution thus comes quite easily as a consequence of the detailed balance condition for an arbitrary choice of the pair (f, q). In practice, the performance of the algorithm will obviously strongly depend on this choice of q, but consider first a straightforward example where Algorithm 4 can be compared with iid sampling.

Example 6.1. Recall Example 2.7, where we used an Accept–Reject algorithm to simulate a beta distribution. We can just as well use a Metropolis–Hastings algorithm, where the target density f is the $\mathcal{Be}(2.7, 6.3)$ density and the candidate q is uniform over $[0, 1]$, which means that it does not depend on the previous value of the chain. A Metropolis–Hastings sample is then generated with the following R code:

```
> a=2.7; b=6.3; c=2.669 # initial values
> Nsim=5000
> X=rep(runif(1),Nsim)   # initialize the chain
> for (i in 2:Nsim){
+    Y=runif(1)
+    rho=dbeta(Y,a,b)/dbeta(X[i-1],a,b)
+    X[i]=X[i-1] + (Y-X[i-1])*(runif(1)<rho)
+    }
```

A representation of the sequence $(X^{(t)})$ by plot does not produce any pattern in the simulation since the chain explores the same range at different periods. If we zoom in on the final period, for $4500 \leq t \leq 4800$, Figure 6.1 exhibits some characteristic features of Metropolis–Hastings sequences, namely that, for some intervals of time, the sequence $(X^{(t)})$ does not change because all corresponding

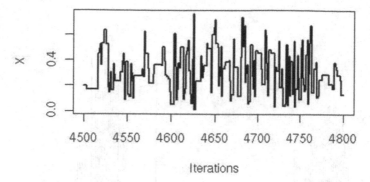

Fig. 6.1. Sequence $X^{(t)}$ for $t = 4500, \dots, 4800$, when simulated from the Metropolis–Hastings algorithm with uniform proposal and $\mathcal{B}e(2.7, 6.3)$ target.

Y_t's are rejected. Note that those multiple occurrences of the same numerical value must be kept in the sample as such; otherwise, the validity of the approximation of f is lost! Indeed, when considering the entire chain as a sample, its histogram properly approximates the $\mathcal{B}e(2.7, 6.3)$ target. Figure 6.2 shows histograms and overlaid densities both for this Metropolis–Hastings sample and for an (exact) iid sample drawn using the rbeta command. The fits are quite similar, and this can be checked even further using a Kolmogorov–Smirnov test of equality between the two samples:

```
> ks.test(jitter(X),rbeta(5000,a,b))

        Two-sample Kolmogorov-Smirnov test

data:  jitter(X) and rbeta(5000,a,b)
D = 0.0202, p-value = 0.2594
alternative hypothesis: two-sided
```

which states that both samples are compatible with the same distribution. An additional (if mild) check of agreement is provided by the moments. For instance, since the mean and variance of a $\mathcal{B}e(a, b)$ distribution are $a/(a + b)$ and $ab/(a + b)^2(a + b + 1)$, respectively, we can compare

$$\bar{X} = .301, \quad S^2 = .0205,$$

with the theoretical values of .3 for the mean and .021 for the variance.   ◄

While the MCMC and exact sampling outcomes look identical in Figure 6.2, it is important to remember that the Markov chain Monte Carlo sample has correlation, while the iid sample does not. This means that the quality of the sample is necessarily degraded or, in other words, that we need more

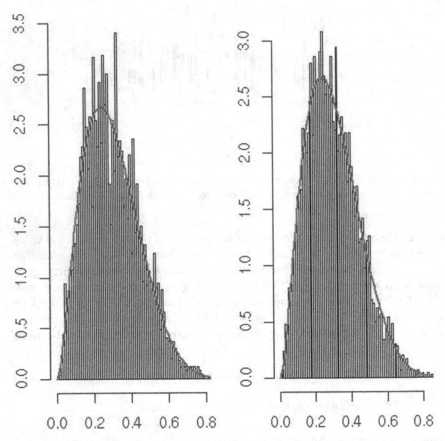

Fig. 6.2. Histograms of beta $\mathcal{B}e(2.7, 6.3)$ random variables with density function overlaid. In the left panel, the variables were generated from a Metropolis–Hastings algorithm with a uniform candidate, and in the right panel the random variables were directly generated using `rbeta(n,2.7,6.3)`.

simulations to achieve the same precision. This issue is formalized through the notion of *effective sample size* for Markov chains (Section 8.4.3).

In the symmetric case (that is, when $q(x|y) = q(y|x)$), the acceptance probability $\rho(x_t, y_t)$ is driven by the objective ratio $f(y_t)/f(x^{(t)})$ and thus even the acceptance probability is independent from q. (This special case is detailed in Section 6.4.1.) Again, Metropolis–Hastings algorithms share the same feature as the stochastic optimization Algorithm 2 (see Section 5.5), namely that they always accept values of y_t such that the ratio $f(y_t)/q(y_t|x^{(t)})$ is increased compared with the "previous" value $f(x^{(t)})/q(x^{(t)}|y_t)$. Some values y_t such that the ratio is decreased may also be accepted, depending on the ratio of the ratios, but if the decrease is too sharp, the proposed value y_t will almost always be rejected. This property indicates how the choice of q can impact the performance of the Metropolis–Hastings algorithm. If the domain explored by q (its *support*) is too small, compared with the range of f, the Markov chain will have difficulties in exploring this range and thus will converge very slowly (if at all for practical purposes).

Another interesting property of the Metropolis–Hastings algorithm that adds to its appeal is that it only depends on the ratios

$$f(y_t)/f(x^{(t)}) \quad \text{and} \quad q(x^{(t)}|y_t)/q(y_t|x^{(t)}).$$

It is therefore independent of normalizing constants. Moreover, since all that matters is the ability to (a) simulate from q and (b) compute the ratio $f(y_t)/q(y_t|x^{(t)})$, q may be chosen in such a way that the intractable parts of f are eliminated in the ratio.

⚡ Since $q(y|x)$ is a conditional density, it integrates to one in y and, as such, involves a functional term that depends on both y and x as well as a normalizing term that depends on x, namely $q(y|x) = C(x)\tilde{q}(x, y)$. When noting above that the Metropolis–Hastings acceptance probability does not depend on normalizing constants, terms like $C(x)$ are obviously excluded from this remark since they must appear in the acceptance probability, lest it jeopardize the stationary distribution of the chain.

6.3.2 The independent Metropolis–Hastings algorithm

The Metropolis–Hastings algorithm of Section 6.3.1 allows a candidate distribution q that only depends on the present state of the chain. If we now require the candidate q to be *independent* of this present state of the chain (that is, $q(y|x) = g(y)$), we do get a special case of the original algorithm:

Algorithm 5 Independent Metropolis–Hastings
Given $x^{(t)}$
    1. Generate $Y_t \sim g(y)$.
    2. Take

$$X^{(t+1)} = \begin{cases} Y_t & \text{with probability} \quad \min\left\{ \dfrac{f(Y_t)\,g(x^{(t)})}{f(x^{(t)})\,g(Y_t)}, 1 \right\} \\ x^{(t)} & \text{otherwise.} \end{cases}$$

This method then appears as a straightforward generalization of the Accept–Reject method in the sense that the instrumental distribution is the same density g as in the Accept–Reject method. Thus, the proposed values Y_t are the same, if not the accepted ones.

Metropolis–Hastings algorithms and Accept–Reject methods (Section 2.3), both being generic simulation methods, have similarities between them that allow comparison, even though it is rather rare to consider using a Metropolis–Hastings solution when an Accept–Reject algorithm is available. In particular, consider that

a. The Accept–Reject sample is iid, while the Metropolis–Hastings sample is not. Although the Y_t's are generated independently, the resulting sample is not iid, if only because the probability of acceptance of Y_t depends on $X^{(t)}$ (except in the trivial case when $f = g$).

b. The Metropolis–Hastings sample will involve repeated occurrences of the same value since rejection of Y_t leads to repetition of $X^{(t)}$ at time $t + 1$. This will have an impact on tests like `ks.test` that do not accept ties.

c. The Accept–Reject acceptance step requires the calculation of the upper bound $M \geq \sup_x f(x)/g(x)$, which is not required by the Metropolis–Hastings algorithm. This is an appealing feature of Metropolis–Hastings if computing M is time-consuming or if the existing M is inaccurate and thus induces a waste of simulations.

Exercise 6.3 Compute the acceptance probability $\rho(x, y)$ in the case $q(y|x) = g(y)$. Deduce that, for a given value $x^{(t)}$, the Metropolis–Hastings algorithm associated with the same pair (f, g) as an Accept–Reject algorithm accepts the proposed value Y_t more often than the Accept–Reject algorithm.

The following exercise gives a first comparison of Metropolis–Hastings with an Accept–Reject algorithm already used in Exercise 2.20 when both algorithms are based on the same candidate.

Exercise 6.4 Consider the target as the $\mathcal{G}(\alpha, \beta)$ distribution and the candidate as the gamma $\mathcal{G}([\alpha], b)$ distribution (where $[a]$ denotes the integer part of a).

a. Derive the corresponding Accept–Reject method and show that, when $\beta = 1$, the optimal choice of b is $b = [\alpha]/\alpha$.

b. Generate 5000 $\mathcal{G}(4, 4/4.85)$ random variables to derive a $\mathcal{G}(4.85, 1)$ sample (note that you will get less than 5000 random variables).

c. Use the same sample in the corresponding Metropolis–Hastings algorithm to generate 5000 $\mathcal{G}(4.85, 1)$ random variables.

d. Compare the algorithms using (i) their acceptance rates and (ii) the estimates of the mean and variance of the $\mathcal{G}(4.85, 1)$ along with their errors. (*Hint:* Examine the correlation in both samples.)

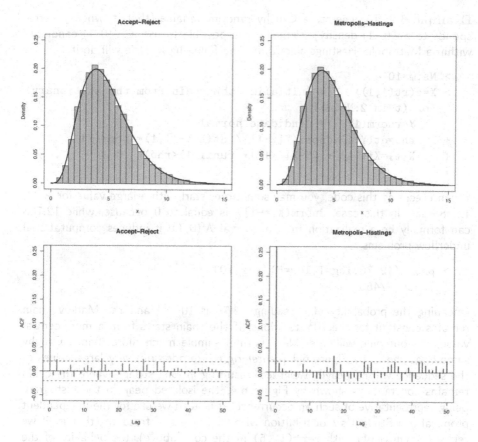

Fig. 6.3. Histograms and autocovariance functions from a gamma Accept–Reject algorithm (left panels) and a gamma Metropolis–Hastings algorithm (right panels). The target is a $\mathcal{G}(4.85, 1)$ distribution and the candidate is a $\mathcal{G}(4, 4/4.85)$ distribution. The autocovariance function is calculated with the R function `acf`.

Figure 6.3 illustrates Exercise 6.4 by comparing both Accept–Reject and Metropolis–Hastings samples. In this setting, operationally, the independent Metropolis–Hastings algorithm performs very similarly to the Accept–Reject algorithm, which in fact generates perfect and independent random variables.

Theoretically, it is also feasible to use a pair (f, g) such that a bound M on f/g does not exist and thus to use Metropolis–Hastings when Accept–Reject is not possible. However, as detailed in Robert and Casella (2004) and illustrated in the following formal example, the performance of the Metropolis–Hastings algorithm is then very poor, while it is very strong as long as $\sup f/g = M < \infty$.

Example 6.2. To generate a Cauchy random variable (that is, when f corresponds to a $\mathcal{C}(0,1)$ density), formally it is possible to use a $\mathcal{N}(0,1)$ candidate within a Metropolis–Hastings algorithm. The following R code will do it:

```
> Nsim=10^4
> X=c(rt(1,1))     # initialize the chain from the stationary
> for (t in 2:Nsim){
+     Y=rnorm(1)    # candidate normal
+     rho=dt(Y,1)*dnorm(X[t-1])/(dt(X[t-1],1)*dnorm(Y))
+     X[t]=X[t-1] + (Y-X[t-1])*(runif(1)<rho)
+     }
```

When executing this code, you may sometimes start with a large value for $X^{(0)}$, 12.788 say. In this case, dnorm(X[t-1]) is equal to 0 because, while 12.788 can formally be a realization from a normal $\mathcal{N}(0,1)$, it induces computational underflow problems

```
> pnorm(12.78,log=T,low=F)/log(10)
[1] -36.97455
```

(meaning the probability of exceeding 12.78 is 10^{-37}) and the Markov chain remains constant for the 10^4 iterations! If the chain starts from a more central value, the outcome will resemble a normal sample much more than a Cauchy sample, as shown by Figure 6.4 *(center right)*. In addition, very large values of the sequence will be heavily weighted, resulting in long strings where the chain remains constant, as shown by Figure 6.4, the isolated peak in the histogram being representative of such an occurrence. If instead we use for the independent proposal g a Student's t distribution with .5 degrees of freedom (that is, if we replace Y=rnorm(1) with Y=rt(1,.5) in the code above), the behavior of the chain is quite different. Very large values of Y_t may occur from time to time (as shown in Figure 6.4 *(upper left)*), the histogram fit is quite good *(center left)*, and the sequence exhibits no visible correlation *(lower left)*. If we consider the approximation of a quantity like $\Pr(X < 3)$, for which the exact value is pt(3,1) (that is, 0.896), the difference between the two choices of g is crystal clear in Figure 6.5, obtained by

```
> plot(cumsum(X<3)/(1:Nsim),lwd=2,ty="l",ylim=c(.85,1)).
```

The chain based on the normal proposal is consistently off the true value, while the chain based on the t distribution with .5 degrees of freedom converges quite quickly to this value. Note that, from a theoretical point of view, the Metropolis–Hastings algorithm associated with the normal proposal still converges, but the convergence is so slow as to be useless.   ◀

We now look at a somewhat more realistic statistical example that corresponds to the general setting when an independent proposal is derived from a preliminary estimation of the parameters of the model. For instance, when simulating from a posterior distribution $\pi(\theta|x) \propto \pi(\theta)f(x|\theta)$, this independent

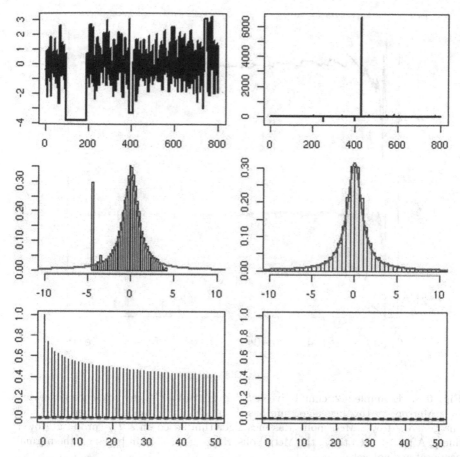

Fig. 6.4. Comparison of two Metropolis–Hastings schemes for a Cauchy target when generating *(left)* from a $\mathcal{N}(0,1)$ proposal and *(right)* from a $\mathcal{T}_{1/2}$ proposal based on 10^5 simulations. *(top)* Excerpt from the chains $(X^{(t)})$; *(center)* histograms of the samples; *(bottom)* autocorrelation graphs obtained by acf.

proposal could be a normal or a t distribution centered at the MLE $\hat{\theta}$ and with variance-covariance matrix equal to the inverse of Fisher's information matrix.

Example 6.3. The cars dataset relates braking distance (y) to speed (x) in a sample of cars. Figure 6.6 shows the data along with a fitted quadratic curve that is given by the R function lm. The model posited for this dataset is a quadratic model

$$y_{ij} = a + bx_i + cx_i^2 + \varepsilon_{ij}, \quad i = 1, \ldots, k, \quad j = 1, \ldots n_i,$$

where we assume that $\varepsilon_{ij} \sim N(0, \sigma^2)$ and independent. The likelihood function is then proportional to

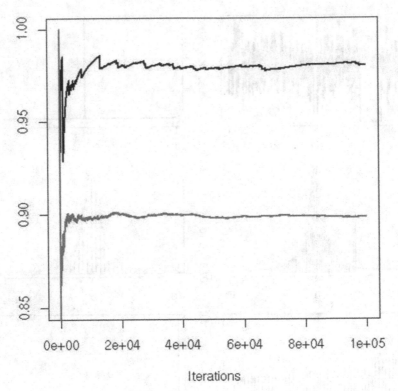

Fig. 6.5. Example 6.2: cumulative coverage plot of a Cauchy sequence generated by a Metropolis–Hastings algorithm based on a $\mathcal{N}(0,1)$ proposal (upper lines) and one generated by a Metropolis–Hastings algorithm based on a $\mathcal{T}_{1/2}$ proposal (lower lines). After 10^5 iterations, the Metropolis–Hastings algorithm based on the normal proposal has not yet converged.

$$\left(\frac{1}{\sigma^2}\right)^{N/2} \exp\left\{\frac{-1}{2\sigma^2}\sum_{ij}(y_{ij} - a - bx_i - cx_i^2)^2\right\},$$

where $N = \sum_i n_i$ is the total number of observations. We can view this likelihood function as a posterior distribution on $a, b, c,$ and σ^2 (for instance based on a flat prior), and, as a toy problem, we can try to sample from this distribution with a Metropolis–Hastings algorithm (since this standard distribution can be simulated directly; see Exercise 6.12). To start with, we can get a candidate by generating coefficients according to their fitted sampling distribution. That is, we can use the R command

```
> x2=x^2
> summary(lm(y~x+x2))
```

to get the output

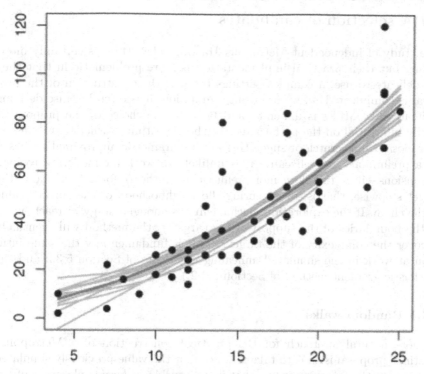

Fig. 6.6. Braking data with quadratic curve *(dark)* fitted with the least squares function lm. The grey curves represent the Monte Carlo sample $(a^{(i)}, b^{(i)}, c^{(i)})$ and show the variability in the fitted lines based on the last 500 iterations of 4000 simulations.

Coefficients:

	Estimate	Std. Error	t value	Pr(> |t|)
(Intercept)	2.63328	14.80693	0.178	0.860
x	0.88770	2.03282	0.437	0.664
x2	0.10068	0.06592	1.527	0.133

Residual standard error: 15.17 on 47 degrees of freedom

As suggested above, we can use the candidate normal distribution centered at the MLEs,

$$a \sim \mathcal{N}(2.63, (14.8)^2), \quad b \sim \mathcal{N}(.887, (2.03)^2), \quad c \sim \mathcal{N}(.100, (0.065)^2),$$

$$\sigma^{-2} \sim \mathcal{G}(n/2, (n-3)(15.17)^2),$$

in a Metropolis–Hastings algorithm to generate samples $(a^{(i)}, b^{(i)}, c^{(i)})$. Figure 6.6 illustrates the variability of the curves associated with the outcome of this simulation.   ◄

6.4 A selection of candidates

The study of independent Metropolis–Hastings algorithms is certainly interesting, but their practical implementation is more problematic in that they
are delicate to use in complex settings because the construction of the proposal is complicated—if we are using simulation, it is often because deriving
estimates like MLEs is difficult—and because the choice of the proposal is
highly influential on the performance of the algorithm. Rather than building
a proposal from scratch or suggesting a non-parametric approximation based
on a preliminary run—because it is unlikely to work for moderate to high
dimensions—it is therefore more realistic to gather information about the
target stepwise, that is, by exploring the neighborhood of the current value
of the chain. If the exploration mechanism has enough energy to reach as far
as the boundaries of the support of the target f, the method will eventually
uncover the complexity of the target. (This is fundamentally the same intuition at work in the simulated annealing algorithm of Section 5.3.3 and the
stochastic gradient method of Section 5.3.2.)

6.4.1 Random walks

A more natural approach for the practical construction of a Metropolis–
Hastings proposal is thus to take into account the value previously simulated
to generate the following value; that is, to consider a *local* exploration of the
neighborhood of the current value of the Markov chain.

The implementation of this idea is to simulate Y_t according to

$$Y_t = X^{(t)} + \varepsilon_t,$$

where ε_t is a random *perturbation* with distribution g independent of $X^{(t)}$,
for instance a uniform distribution or a normal distribution, meaning that
$Y_t \sim \mathcal{U}(X^{(t)} - \delta, X^{(t)} + \delta)$ or $Y_t \sim \mathcal{N}(X^{(t)}, \tau^2)$ in unidimensional settings.
In terms of the general Metropolis–Hastings algorithm, the proposal density
$q(y|x)$ is now of the form $g(y - x)$. The Markov chain associated with q is
a *random walk* (as described in Section 6.2) when the density g is symmetric around zero; that is, satisfying $g(-t) = g(t)$. But, due to the additional
Metropolis–Hastings acceptance step, the Metropolis–Hastings Markov chain
$\{X^{(t)}\}$ is *not* a random walk. This approach leads to the following Metropolis–
Hastings algorithm, which also happens to be the original one proposed by
Metropolis et al. (1953).

Algorithm 6 Random walk Metropolis–Hastings
Given $x^{(t)}$,
    1. Generate $Y_t \sim g(y - x^{(t)})$.

---

2. Take

$$X^{(t+1)} = \begin{cases} Y_t & \text{with probability } \min\left\{1, f(Y_t)/f(x^{(t)})\right\}, \\ x^{(t)} & \text{otherwise.} \end{cases}$$

As noted above, the acceptance probability does not depend on g. This means that, for a given pair $(x^{(t)}, y_t)$, the probability of acceptance is the same whether y_t is generated from a normal or from a Cauchy distribution. Obviously, changing g will result in different ranges of values for the Y_t's and a different acceptance rate, so this is not to say that the choice of g has no impact whatsoever on the behavior of the algorithm, but this invariance of the acceptance probability is worth noting. It is actually linked to the fact that, for *any* (symmetric) density g, the invariant measure associated with the random walk is the Lebesgue measure on the corresponding space (see Meyn and Tweedie, 1993).

Example 6.4. The historical example of Hastings (1970) considers the formal problem of generating the normal distribution $\mathcal{N}(0,1)$ based on a random walk proposal equal to the uniform distribution on $[-\delta, \delta]$. The probability of acceptance is then

$$\rho(x^{(t)}, y_t) = \exp\{(x^{(t)2} - y_t^2)/2\} \wedge 1.$$

Figure 6.7 describes three samples of 5000 points produced by this method for $\delta = 0.1, 1,$ and 10 and clearly shows the difference in the produced chains: Too narrow or too wide a candidate (that is, a smaller or a larger value of δ) results in higher autocovariance and slower convergence. Note the distinct patterns for $\delta = 0.1$ and $\delta = 10$ in the upper graphs: In the former case, the Markov chain moves at each iteration but very slowly, while in the latter it remains constant over long periods of time.   ◀

As noted in this formal example, calibrating the scale δ of the random walk is crucial to achieving a good approximation to the target distribution in a reasonable number of iterations. In more realistic situations, this calibration becomes a challenging issue, partly tackled in Section 6.5 and reconsidered in further detail in Chapter 8.

Example 6.5. The mixture example detailed in Example 5.2 from the perspective of a maximum likelihood estimation can also be considered from a Bayesian point of view using for instance a uniform prior $\mathcal{U}(-2,5)$ on both μ_1 and μ_2. The posterior distribution we are interested in is then proportional to the likelihood. Implementing Algorithm 6 in this example is surprisingly easy in that we can recycle most of the implementation of the simulated annealing Algorithm 2, already

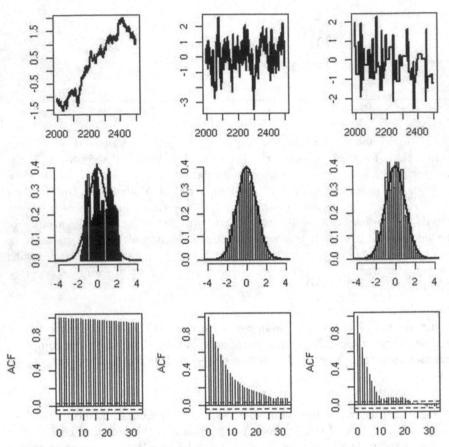

Fig. 6.7. Outcomes of random walk Metropolis–Hastings algorithms for Example 6.4. The left panel has a $\mathcal{U}(-.1, .1)$ candidate, the middle panel has $\mathcal{U}(-1, 1)$, and the right panel has $\mathcal{U}(-10, 10)$. The upper graphs represent the last 500 iterations of the chains, the middle graphs indicate how the histograms fit the target, and the lower graphs give the respective autocovariance functions.

programmed in Example 5.2. Indeed, the core of the R code is very similar except for the increase in temperature, which obviously is not necessary here:

```
> scale=1
> the=matrix(runif(2,-2,5),ncol=2)
> curlike=hval=like(x)
> Niter=10^4
> for (iter in (1:Niter)){
+    prop=the[iter,]+rnorm(2)*scale
+    if ((max(-prop)>2)||(max(prop)>5)||
+       (log(runif(1))>like(prop)-curlike)) prop=the[iter,]
```

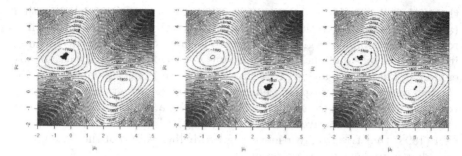

Fig. 6.8. Impact of the scale of the random walk on the exploration of the modes in the mixture model: representation of the Markov chain $(\mu_1^{(t)}, \mu_2^{(t)})$ on top of the log posterior surface with *(left and center)* scale equal to 1 and *(right)* scale equal to 2 based on 10^4 simulations and 500 simulated observations.

```
+    curlike=like(prop)
+    hval=c(hval,curlike)
+    the=rbind(the,prop)}
```

Since the main problem of this target is the existence of two modes, one of which is smaller than the other, we can compare the impact of different choices of scale on the behavior of the chain in terms of exploration of both modes and the attraction therein. When the scale is 1, the modes are highly attractive and, out of 10^4 iterations, it is not uncommon to explore only one mode neighborhood, as shown in Figure 6.8 (left and center) for both modes. If the scale increases to 2, the proposal is diverse enough to reach both modes but at a cost. Out of 10^4 iterations, the chain only changes values 23 times! For the smaller scale 1, the number of changes is closer to 100, still a very low acceptance rate.   ◄

An issue that often arises when using random walks on constrained domains is whether or not the random walk should be constrained as well. The answer to this question is no in that using constraints in the proposal modifies the function g and thus jeopardizes the validity of the ratio of the targets found in Algorithm 6. When values y_t outside the range of f are proposed (that is, when $f(y_t) = 0$), the proposed value is rejected and the current value $X^{(t)}$ is duplicated. Obviously, picking a random walk density that often ends up outside the domain of f is a poor idea in that the chain will be stuck most of the time! But it is formally correct.

6.4.2 Alternative candidates

While the independent Metropolis–Hastings algorithm only applies in specific situations, the random walk Metropolis–Hastings algorithm often appears as a generic Metropolis–Hastings algorithm that caters to most cases. Nonetheless, the random walk solution is not necessarily the most efficient choice in that (a) it requires many iterations to overcome difficulties such as low-probability regions between modal regions of f and (b) because of its symmetric features, it spends roughly half the simulation time revisiting regions it has already explored. There exist alternatives that bypass the perfect symmetry in the random walk proposal to gain in efficiency, although they are not always easy to implement (see, for example, Robert and Casella, 2004).

One of those alternatives is the Langevin algorithm of Roberts and Rosenthal (1998) that tries to favor moves toward higher values of the target f by including a gradient in the proposal,

$$Y_t = X^{(t)} + \frac{\sigma^2}{2} \nabla \log f(X^{(t)}) + \sigma \epsilon_t, \qquad \epsilon_t \sim g(\epsilon),$$

the parameter σ being the scale factor of the proposal. When Y_t is constructed this way, the Metropolis–Hastings acceptance probability is equal to

$$\rho(x,y) = \min \left\{ \frac{f(y)}{f(x)} \frac{g\left[(x-y)/\sigma - \sigma \nabla \log f(y)/2\right]}{g\left[(y-x)/\sigma - \sigma \nabla \log f(x)/2\right]}, 1 \right\}.$$

While this scheme may remind you of the stochastic gradient techniques of Section 5.3.2, it differs from those for two reasons. One is that the scale σ is fixed in the Langevin algorithm, as opposed to decreasing in the stochastic gradient method. Another is that the proposed move to Y_t is not necessarily accepted for the Langevin algorithm, ensuring the stationarity of f for the resulting chain.

Example 6.6. Based on the same probit model of the now well-known Pima.tr dataset as in Example 3.10, we can use the likelihood function like already defined on page 85 and compute the gradient in closed form as

```
grad=function(a,b){
    don=pnorm(q=a+outer(X=b,Y=da[,2],FUN="*"))
    x1=sum(((dnorm(x=a+outer(X=b,Y=da[,2],FUN="*"))/don)*da[,1]-
            (dnorm(x=-a-outer(X=b,Y=da[,2],FUN="*"))/
                (1-don))*(1-da[,1]))
    x2=sum(da[,2]*(
        (dnorm(x=a+outer(X=b,Y=da[,2],FUN="*"))/don)*da[,1]-
        (dnorm(x=-a-outer(X=b,Y=da[,2],FUN="*"))/
            (1-don))*(1-da[,1])))
    return(c(x1,x2))
    }
```

When implementing the basic iteration of the Langevin algorithm

```
>   prop=curmean+scale*rnorm(2)
>   propmean=prop+0.5*scale^2*grad(prop[1],prop[2])
```

```
>  if (log(runif(1))>like(prop[1],prop[2])-likecur-
+     sum(dnorm(prop,mean=curmean,sd=scale,lo=T))+
+     sum(dnorm(the[t-1,],mean=propmean,sd=scale,lo=T)))){
+        prop=the[t-1,];propmean=curmean}
```

we need to select scale small enough because otherwise grad(prop) returns NaN
given that pnorm(q=a+outer(X=b,Y=da[,2],FUN="*")) is then either 1 or 0.
With a scale equal to 0.01, the chain correctly explores the posterior distribution,
as shown in Figure 6.9, even though it moves very slowly.   ◄

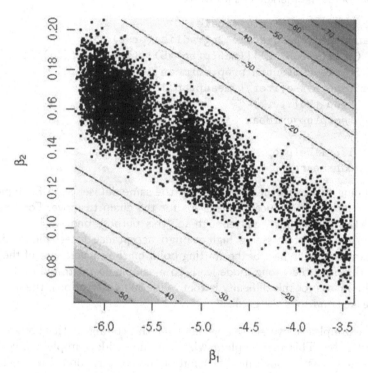

Fig. 6.9. Repartition of the Langevin sample corresponding to the probit posterior
defined in Example 3.10 based on 20 observations from Pima.tr and 5×10^4 iterations.

The modification of the random walk proposal may, however, further hinder the mobility of the Markov chain by reinforcing the polarization around
local modes. For instance, when the target is the posterior distribution of the
mixture model studied in Example 6.5, the bimodal structure of the target
is a hindrance for the implementation of the Langevin algorithm in that the
local mode becomes even more attractive.

Example 6.7. (Continuation of Example 6.5) The modification of the random walk Metropolis–Hastings algorithm is straightforward in that we simply have to add the gradient drift in the R code. Defining the gradient function

```
gradlike=function(mu){
  deno=.2*dnorm(da-mu[1])+.8*dnorm(da-mu[2])
  gra=sum(.2*(da-mu[1])*dnorm(da-mu[1])/deno)
  grb=sum(.8*(da-mu[2])*dnorm(da-mu[2])/deno)
  return(c(gra,grb))
  }
```

the simulation of the Markov chain involves

```
> prop=curmean+rnorm(2)*scale
> meanprop=prop+.5*scale^2*gradlike(prop)
> if ((max(-prop)>2)||(max(prop)>5)||(log(runif(1))>like(prop)
+   -curlike-sum(dnorm(prop,curmean,lo=T))+
+   sum(dnorm(the[iter,],meanprop,lo=T)))){
+      prop=the[iter,]
+      meanprop=curmean
+      }
> curlike=like(prop)
> curmean=meanprop
```

When running this Langevin alternative on the same dataset as in Example 6.5, the scale needs to be reduced quite a lot for the chain to move. For instance, using scale=.2 was not small enough for this purpose and we had to lower it to scale=.1 to start seeing high enough acceptance rates. Figure 6.10 is representative of the impact of the starting point on the convergence of the chain since starting near the wrong mode leads to a sample concentrated on this very mode. The reason for this difficulty is that, with 500 observations, the likelihood is very peaked and so is the gradient.   ◀

Both examples above show how delicate the tuning of the Langevin algorithm can be. This may explain why it is not widely implemented, even though it is an easy enough modification of the basic random walk code.

Random walk Metropolis–Hastings algorithms also apply to discrete support targets. While this sounds more like a combinatoric or an image-processing setting, since most statistical problems involve continuous parameter spaces, an exception is the case of model choice (see, for example, Robert, 2001, Chapter 7), where the index of the model to be selected is the "parameter" of interest.

Example 6.8. Given an ordinary linear regression with n observations,

$$\mathbf{y}|\beta,\sigma^2,X \sim \mathcal{N}_n(X\beta,\sigma^2 I_n),$$

where X is a (n,p) matrix, the likelihood is

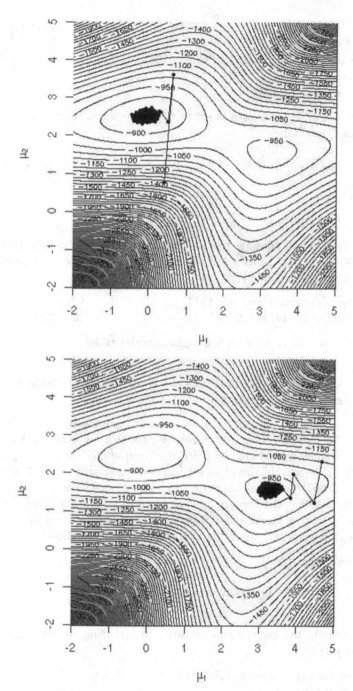

Fig. 6.10. Exploration of the modes in the mixture model by a Langevin algorithm: representation of two Markov chains $(\mu_1^{(t)}, \mu_2^{(t)})$ on top of the log-posterior surface with a scale equal to .1 based on 10^4 simulations and a simulated dataset of 500 observations.

$$\ell\left(\beta,\sigma^2|\mathbf{y},X\right) = \left(2\pi\sigma^2\right)^{-n/2}\exp\left[-\frac{1}{2\sigma^2}(\mathbf{y}-X\beta)^{\mathsf{T}}(\mathbf{y}-X\beta)\right]$$

and, under the so-called g-prior of Zellner (1986),

$$\beta|\sigma^2,X \sim \mathcal{N}_{k+1}(\tilde{\beta},n\sigma^2(X^{\mathsf{T}}X)^{-1}) \quad \text{and} \quad \pi(\sigma^2|X) \propto \sigma^{-2}$$

(where the constant g is chosen equal to n), the marginal distribution of $\mathbf{y}$ is a multivariate t distribution,

$$m(\mathbf{y}|X) = (n+1)^{-(k+1)/2}\pi^{-n/2}\Gamma(n/2)\left[\mathbf{y}^{\mathsf{T}}\mathbf{y} - \frac{n}{n+1}\mathbf{y}^{\mathsf{T}}X(X^{\mathsf{T}}X)^{-1}X^{\mathsf{T}}\mathbf{y}\right.$$
$$\left. - \frac{1}{n+1}\tilde{\beta}^{\mathsf{T}}X^{\mathsf{T}}X\tilde{\beta}\right]^{-n/2}.$$

As an illustration, we consider the swiss dataset, where the logarithm of the fertility in 47 districts of Switzerland around 1888 is the variable y to be explained by some socioeconomic indicators,

```
> y=log(as.vector(swiss[,1]))
> X=as.matrix(swiss[,2:6])
```

The covariate matrix X involves five explanatory variables

```
> names(swiss)
[1] "Fertility"  "Agriculture"  "Examination"  "Education"
[5] "Catholic"   "Infant.Mortality"
```

(that are explained by ?swiss) and we want to compare the 2^5 models corresponding to all possible subsets of covariates. (In this toy example, the number of models is small enough to allow for the computation of all marginals and therefore the true probabilities of all models under comparison.) Following Marin and Robert (2007), we index all models by vectors γ of binary indicators where $\gamma_i = 0$ indicates that the corresponding column of X is used in the regression. (Note that, adopting Marin and Robert's, 2007, convention, we always include the intercept in a model.) Using the fast inverse matrix function

```
inv=function(X){
  EV=eigen(X)
  EV$vector%*%diag(1/EV$values)%*%t(EV$vector)
  }
```

we then compute the log marginal density corresponding to the model γ, now denoted as $m(\mathbf{y}|X,\gamma)$, as

```
lpostw=function(gam,y,X,beta){
  n=length(y)
  qgam=sum(gam)
  Xt1=cbind(rep(1,n),X[,which(gam==1)])
```

```
   if (qgam!=0) P1=Xt1%*%inv(t(Xt1)%*%Xt1)%*%t(Xt1) else{
      P1=matrix(0,n,n)}
      -(qgam+1)/2*log(n+1)-n/2*log(t(y)%*%y-n/(n+1)*
      t(y)%*%P1%*%y-1/(n+1)*t(beta)%*%t(cbind(rep(1,n),
      X))%*%P1%*%cbind(rep(1,n),X)%*%beta)
   }
```

The exploration of the space of models can result from a Metropolis–Hastings algorithm that moves around models by changing one model indicator at a time; that is, given the current indicator vector $\gamma^{(t)}$, the Metropolis–Hastings proposal picks one of the p coordinates, say i, and chooses between keeping $\gamma_i^{(t)}$ and switching to $1 - \gamma_i^{(t)}$ with probabilities proportional to the associated marginals. The Metropolis–Hastings acceptance probability of the proposed model $\gamma^\star$ is then equal to

$$\min\left\{\frac{m(\mathbf{y}|X,\gamma^\star)}{m(\mathbf{y}|X,\gamma^{(t)})}\,\frac{m(\mathbf{y}|X,\gamma^{(t)})}{m(\mathbf{y}|X,\gamma^\star)},1\right\} = 1$$

since the normalising constants cancel. This means that we do not have to consider rejecting the proposed model $\gamma^\star$ because it is always accepted at the Metropolis–Hastings step! Running the R function

```
gocho=function(niter,y,X){
   lga=dim(X)[2]
   beta=lm(y~X)$coeff
   gamma=matrix(0,nrow=niter,ncol=lga)
   gamma[1,]=sample(c(0,1),lga,rep=T)
   for (t in 1:(niter-1)){
      j=sample(1:lga,1)
      gam0=gam1=gamma[t,];gam1[j]=1-gam0[j]
      pr=lpostw(gam0,y,X,beta)
      pr=c(pr,lpostw(gam1,y,X,beta))
      pr=exp(pr-max(pr))
      gamma[t+1,]=gam0
      if (sample(c(0,1),1,prob=pr))
         gamma[t+1,]=gam1}
   gamma
   }
```

then produces a sample (approximately) distributed from the posterior distribution on the set of indicators; that is, on the collection of possible submodels. Based on the outcome

```
> out=gocho(10^5,y,X)
```

the most likely model corresponds to the exclusion of the Agriculture variable (that is, $\gamma = (1,0,1,1,1)$), with estimated probability 0.4995, while the true probability is 0.4997. (This model is also the one indicated by lm(y~X).) Similarly, the second most likely model is $\gamma = (0, 0, 1, 1, 1)$, with an estimated probability of 0.237 versus a true probability of 0.234. The probability that each variable is included within the model is also provided by

```
> apply(out,2,mean)
[1] 0.66592 0.17978 0.99993 0.91664 0.94499
```

which, again, indicates that the last three variables of swiss are the most significant in this analysis.   ◀

The fact that the acceptance probability is always equal to 1 in Example 6.8 is due to the use of the true target probability on a subset of the possible values of the model indicator.

Exercise 6.5 Starting from the prior distribution

$$\beta | \sigma^2, X \sim \mathcal{N}_{k+1}(\tilde{\beta}, n\sigma^2(X^\mathsf{T} X)^{-1}) :$$

a. Show that

$$X\beta | \sigma^2, X \sim \mathcal{N}_n(X\tilde{\beta}, n\sigma^2 X(X^\mathsf{T} X)^{-1} X^\mathsf{T})$$

and that

$$y | \sigma^2, X \sim \mathcal{N}_n(X\tilde{\beta}, \sigma^2(I_n + nX(X^\mathsf{T} X)^{-1} X^\mathsf{T})) .$$

b. Show that integrating in σ^2 with $\pi(\sigma^2) = 1/\sigma^2$ yields the marginal distribution of y above.
c. Compute the value of the marginal density of y for the swiss dataset.

6.5 Acceptance rates

There are infinite choices for the candidate distribution q in a Metropolis–Hastings algorithm, and here we discuss the possibility of achieving an "optimal" choice. Most obviously, this is not a well-defined concept in that the "optimal" choice of q is to take $q = f$, the target distribution, when reasoning in terms of speed of convergence. This is obviously a formal result that has no relevance in practice! Instead, we need to adopt a practical criterion that allows the comparison of proposal kernels in situations where (almost) nothing is known about f. One such criterion is the acceptance rate of the corresponding Metropolis–Hastings algorithm since it can be easily computed when running this algorithm via the empirical frequency of acceptance. In contrast to Chapter 2, where the calibration of an Accept–Reject algorithm was based on the maximum acceptance rate, merely optimizing the acceptance rate will not necessarily result in the best algorithm in terms of mixing and convergence.

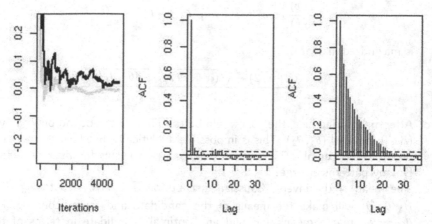

Fig. 6.11. Cumulative mean plot (left) from a Metropolis–Hastings algorithm used to generate a $\mathcal{N}(0,1)$ random variable from a double-exponential proposal distribution $\mathcal{L}(1)$ (lighter) and $\mathcal{L}(3)$ (black). The center and left panels show the autocovariance for the $\mathcal{L}(1)$ and $\mathcal{L}(3)$ proposals, respectively.

Example 6.9. In an Accept–Reject algorithm generating a $\mathcal{N}(0,1)$ sample from a double-exponential distribution $\mathcal{L}(\alpha)$ with density $g(x|\alpha) = (\alpha/2)\exp(-\alpha|x|)$, the choice $\alpha = 1$ optimizes the acceptance rate (Exercise 2.19). We can use this distribution as an independent candidate q in a Metropolis–Hastings algorithm. Figure 6.11 compares the behavior of this $\mathcal{L}(1)$ candidate along with an $\mathcal{L}(3)$ distribution, which, for this simulation, produces an inferior outcome in the sense that it has larger autocovariances and, as a result of this, slower convergence. Obviously, a deeper analysis would be necessary to validate this statement, but our point here is that the acceptance rate (estimated) for $\alpha = 1$ is twice as large, 0.83, as the acceptance rate (estimated) for $\alpha = 3$, 0.47. ◀

While independent Metropolis–Hastings algorithms can indeed be optimized or at least compared through their acceptance rate, because this reduces the number of replicas in the chain $\{X^{(t)}\}$ and thus the correlation level in the chain, this is not true for other types of Metropolis–Hastings algorithms, first and foremost the random walk version.

Exercise 6.6 The *inverse Gaussian distribution* has the density

$$f(z|\theta_1, \theta_2) \propto z^{-3/2} \exp\left\{-\theta_1 z - \frac{\theta_2}{z} + 2\sqrt{\theta_1\theta_2} + \log\sqrt{2\theta_2}\right\}$$

on $\mathbb{R}_+$ ($\theta_1 > 0, \theta_2 > 0$).

a. A candidate for a Metropolis–Hastings algorithm targeting f is the $\mathcal{G}(\alpha, \beta)$ distribution. Show that

$$\frac{f(x)}{g(x)} \propto x^{-\alpha-1/2} \exp\left\{(\beta - \theta_1)x - \frac{\theta_2}{x}\right\}$$

is maximized in x at

$$x_\beta^* = \frac{(\alpha + 1/2) - \sqrt{(\alpha + 1/2)^2 + 4\theta_2(\theta_1 - \beta)}}{2(\beta - \theta_1)}.$$

b. After maximizing in x, the goal would be to minimize the bound on f/g over (α, β) for fixed (θ_1, θ_2). This is impossible analytically, but for chosen values of (θ_1, θ_2) we can plot this function of (α, β). Do so using for instance persp. Do any patterns emerge?

c. The mean of the inverse Gaussian distribution is $\sqrt{\theta_2/\theta_1}$, so taking $\alpha = \beta\sqrt{\theta_2/\theta_1}$ will make the means of the candidate and target coincide. For $\theta_1 = \theta_2$, match means and find an "optimal" candidate in terms of the acceptance rate.

The *random walk* version of the Metropolis–Hastings algorithm, introduced in Section 6.4.1, does indeed require a different approach to acceptance rates, given the dependence of the candidate distribution on the current state of the chain. In fact, as already seen in Example 6.4, a high acceptance rate does not necessarily indicate that the algorithm is behaving satisfactorily since it may instead correspond to the fact that the chain is moving too slowly on the surface of f. When $x^{(t)}$ and y_t are close, in the sense that $f(x^{(t)})$ and $f(y_t)$ are approximately equal, the random walk Metropolis–Hastings algorithm leads to the acceptance of y_t with probability

$$\min\left(\frac{f(y_t)}{f(x^{(t)})}, 1\right) \simeq 1.$$

A high acceptance rate may therefore signal a poor convergence pattern as the moves on the support of f are more limited. Obviously, this is not always the case. For instance, when f is nearly flat, high acceptance rates are not indicative of any wrong behavior! But, unless f is completely flat (that is, it corresponds to a uniform target), there are parts of the domain to be explored where f takes smaller values and hence where the acceptance probabilities should be small. A high acceptance rate then indicates that those parts of the domain are not often (or not at all!) explored by the Metropolis–Hastings algorithm.

In contrast, if the average acceptance rate is low, the successive values of $f(y_t)$ often are small when compared with $f(x^{(t)})$, which corresponds to the scenario where the random walk moves quickly on the surface of f since it often reaches the "borders" of the support of f (or at least when the random walk explores regions with low probability under f). Again, a low acceptance rate does not mean that the chain explores the entire support of f. Even with a small acceptance rate, it may miss an important but isolated mode of f. Nonetheless, a low acceptance rate is less of an issue, except from the computing time point of view, because it explicitly indicates that a larger number of simulations are necessary. Using the effective sample size as a convergence indicator (see Section 8.4.3) would clearly signal this requirement.

Example 6.10. (Continuation of Example 6.4) The three random walk Metropolis–Hastings algorithms of Figure 6.7 have acceptance rates equal to

[1] 0.9832
[1] 0.7952
[1] 0.1512

respectively. Looking at the histogram fit, we see that the medium acceptance rate does better but that the lowest acceptance rate still fares better than the highest one.   ◀

The question is then to decide on a golden acceptance rate against which to calibrate random walk Metropolis–Hastings algorithms in order to avoid "too high" as well as "too low" acceptance rates. Roberts et al. (1997) recommend the use of instrumental distributions with *acceptance rates close to* 1/4 *for models of high dimension and equal to* 1/2 *for the models of dimension* 1 *or* 2. (This is the rule adopted in the adaptive amcmc package described in Section 8.5.2.) While this rule is not universal (in the sense that it was primarily designed for a Gaussian environment), we advocate it as a default calibration goal whenever it can be achieved (which is not always the case). For instance, if we consider the Metropolis–Hastings algorithm in Example 6.8, there is *no* acceptance rate since the acceptance probability is always equal to 1. However, since the proposal includes the current value in its support, the chain $\{\gamma^{(t)}\}$ has identical values in a row and thus an implicit acceptance (or renewal) rate. It is equal to 0.1805, much below the 0.25 goal, and the algorithm cannot be easily modified (for instance, by looking at more alternative moves around the current model) to reach this token acceptance rate.

6.6 Additional exercises

Exercise 6.7 Referring to Example 2.7, consider a $\mathcal{Be}(2.7, 6.3)$ target density.

a. Generate Metropolis–Hastings samples from this density using a range of independent beta candidates from a $\mathcal{Be}(1, 1)$ to a beta distribution with small variance. (*Note:* Recall that the variance is $ab/(a+b)^2(a+b+1)$.) Compare the acceptance rates of the algorithms.

b. Suppose that we want to generate a *truncated* beta $\mathcal{Be}(2.7, 6.3)$ restricted to the interval (c, d) with $c, d \in (0, 1)$. Compare the performance of a Metropolis–Hastings algorithm based on a $\mathcal{Be}(2, 6)$ proposal with one based on a $\mathcal{U}(c, d)$ proposal. Take $c = .1, .25$ and $d = .9, .75$.

Exercise 6.8 While q is a Markov kernel used in Algorithm 4, it is not the Markov kernel K of the algorithm.

1. Show that the probability that $X^{(t+1)} = x^{(t)}$ is

$$\underline{\rho}(x^{(t)}) = \int \left\{ 1 - \rho(x^{(t)}, y) \right\} q(y|x^{(t)}) \, dy \, .$$

2. Deduce that the kernel K can be written as

$$K(x^{(t)}, y) = \rho(x^{(t)}, y) q(y|x^{(t)}) + \underline{\rho}(x^{(t)}) \delta_{x^{(t)}}(y) \, .$$

3. Show that Algorithm 4 satisfies the *detailed balance condition* (6.3).

Exercise 6.9 Calculate the mean of a gamma $\mathcal{G}(4.3, 6.2)$ random variable using

a. Accept–Reject with a gamma $\mathcal{G}(4, 7)$ candidate;
b. Metropolis–Hastings with a gamma $\mathcal{G}(4, 7)$ candidate;
c. Metropolis–Hastings with a gamma $\mathcal{G}(5, 6)$ candidate.

In each case, monitor the convergence across iterations.

Exercise 6.10 Student's t density with ν degrees of freedom, $\mathcal{T}_\nu$, is given by

$$f(x|\nu) = \frac{\Gamma\left(\frac{\nu+1}{2}\right)}{\Gamma\left(\frac{\nu}{2}\right)} \frac{1}{\sqrt{\nu\pi}} \left(1 + x^2/\nu\right)^{-(\nu+1)/2} \, .$$

Calculate the mean of a t distribution with $\nu = 4$ degrees of freedom using a Metropolis–Hastings algorithm with candidate density

a. $\mathcal{N}(0, 1)$;
b. t with $\nu = 2$ degrees of freedom.

In each case monitor the convergence across iterations.

Exercise 6.11 Referring to Example 6.3:

1. Use the candidate given in this example to generate a sample $(a^{(i)}, b^{(i)}, c^{(i)})$, $i = 1, \ldots, 500$ with a Metropolis–Hastings algorithm. The data is from the dataset cars.
2. Monitor convergence and check autocorrelations for each parameter across iterations.
3. Make histograms of the posterior distributions of the coefficient estimates, and provide 95% confidence intervals.

Exercise 6.12 Still in connection with Example 6.3, show that the posterior distribution on (a, b, c, σ^{-2}) is a standard distribution made of a trivariate normal on (a, b, c) conditional on σ and the data and a gamma distribution on σ^{-2} given the data. (*Hint:* See Robert, 2001, or Marin and Robert, 2007, for details.)

Exercise 6.13 In 1986, the space shuttle Challenger exploded during takeoff, killing the seven astronauts aboard. The explosion was the result of an *O-ring* failure, a splitting of a ring of rubber that seals the parts of the ship together. The accident was believed to have been caused by the unusually cold weather ($31°$ F or $0°$ C) at the time of launch, as there is reason to believe that the O-ring failure probabilities increase as temperature decreases. Data on previous space shuttle launches and O-ring failures is given in the dataset challenger provided with the mcsm package. The first column corresponds to the failure indicators y_i and the second column to the corresponding temperature x_i ($1 \le i \le 24$).

1. Fit this dataset with a logistic regression, where

$$P(Y_i = 1|x_i) = p(x_i) = \exp(\alpha + \beta x_i)/1 + \exp(\alpha + \beta x_i),$$

   using R glm function, as illustrated on page 21. Deduce the MLEs for α and β, along with standard errors.
2. Set up a Metropolis–Hastings algorithm with the likelihood as target using an exponential candidate for α and a Laplace (double-exponential) candidate for β. (*Hint:* Choose the parameters of the candidates based on the MLEs derived in a.)
3. Generate 5000 iterations of the Markov chain and construct a picture similar to Figure 6.6 to evaluate the variability of $p(x)$ minus the observation dots.
4. Derive from this sample an estimate of the probability of failure at $60°$, $50°$, and $40°$ F along with a standard error.

Exercise 6.14 Referring to Example 6.4:

a. Reproduce the graphs in Figure 6.7 for different values of δ. Explore both small and large δ's. Can you find an optimal choice in terms of autocovariance?
b. The random walk candidate can be based on other distributions. Consider generating a $\mathcal{N}(0,1)$ distribution using a random walk with a (*i*) Cauchy candidate, and a (*ii*) Laplace candidate. Construct these Metropolis–Hastings algorithms and compare them with each other and with the Metropolis–Hastings random walk with a uniform candidate.
c. For each of these three random walk candidates, examine whether or not the acceptance rate can be brought close to 0.25 for the proper choice of parameters.

Exercise 6.15 Referring to Example 6.9:

a. Write a Metropolis–Hastings algorithm to produce Figure 6.11. Note that n $\mathcal{L}(a)$ random variables can be generated at once with the R command
   ```
   > ifelse(runif(n)>0.5, 1, -1) * rexp(n)/a
   ```
b. What is the acceptance rate for the Metropolis–Hastings algorithm with candidate $\mathcal{L}(3)$? Plot the curve of the acceptance rates for $\mathcal{L}(\alpha)$ candidates when α varies between 1 and 10. Comment.
c. Plot the curve of the acceptance rates for candidates $\mathcal{L}(0,\omega)$ when ω varies between .01 and 10. Compare it with those of the $\mathcal{L}(\alpha)$ candidates.
d. Plot the curve of the acceptance rates when the proposal is based on a random walk, $Y = X^{(t)} + \varepsilon$, where $\varepsilon \sim \mathcal{L}(\alpha)$. Once again, compare it with the earlier proposals.

Exercise 6.16 In connection with Example 6.8, compare the current implementation with an alternative where more values are considered at once according to the R code

```
> progam=matrix(gama[i,],ncol=lga,nrow=lga,byrow=T)
> probam=rep(0,lga)
> for (j in 1:lga){
+    progam[j,j]=1-gama[i,j]
+    probam[j]=lpostw(progam[j,],y,X,betatilde)}
> probam=exp(probam)
> sumam=sum(probam)
> probam=probam/sumam
> select=progam[sample(1:lga,1,prob=probam),]
```

a. Show that the acceptance probability is different from 1 and involves sumam.
b. Study the speed of convergence of the evaluation of the posterior probability of the most likely model in comparison with the implementation on page 191.

7

Gibbs Samplers

"Come, Watson , come!" he cried. "The game is afoot."

Arthur Conan Doyle
The Adventure of the Abbey Grange

Reader's guide

This chapter covers both the two-stage and the multistage Gibbs samplers. Although the former is a special case of the latter, the two-stage sampler has superior convergence properties and applies naturally in a wide range of statistical models that do not call for the generality of the multistage sampler. Nevertheless, the multistage Gibbs sampler enjoys many optimality properties and still might be considered the workhorse of the MCMC world. Following the introduction in Section 7.1 with some background, we develop the two-stage Gibbs sampler in Section 7.2, moving to the multistage Gibbs sampler in Section 7.3. The Gibbs sampler is particularly well-suited to handle experiments with missing data and models with latent variables, as shown in Section 7.4. Although we make use of hierarchical models throughout the chapter, we focus on their processing in Section 7.5. Section 7.6 looks at a number of additional topics such as Rao–Blackwellization, reparameterization, and the effect of using improper priors.

C.P. Robert, G. Casella, *Introducing Monte Carlo Methods with R*, Use R,
DOI 10.1007/978-1-4419-1576-4_7, © Springer Science+Business Media, LLC 2010

7.1 Introduction

Chapter 6 described some principles for simulation based on Markov chains, as well as some implementation directions, including the generic random walk Metropolis–Hastings algorithm. This chapter extends the scope of MCMC algorithms by studying another class of now-common MCMC methods, called Gibbs sampling. The appeal of those specific algorithms is that first they gather most of their calibration from the target density and second they allow us to break complex problems (such as high dimensional target distributions, for which a random walk Metropolis–Hastings algorithm is almost impossible to build) into a series of easier problems, like a sequence of small-dimension targets. There may be caveats to this simplification in that the sequence of simple problems may take *in fine* a long time to converge, but Gibbs sampling is nonetheless an interesting candidate when dealing with a new problem.

The name *Gibbs sampling* comes from the landmark paper by Geman and Geman (1984), which first applied a Gibbs sampler on a *Gibbs random field*. For good or bad, it then stuck despite this weak link. Indeed, it is in fact a special case of the Metropolis–Hastings algorithm as detailed in Robert and Casella (2004, Section 10.6.1). The work of Geman and Geman (1984), built on that of Metropolis et al. (1953), Hastings (1970) and Peskun (1973), influenced Gelfand and Smith (1990) to write a paper that sparked new interest in Bayesian methods, statistical computing, algorithms, and stochastic processes through the use of computing algorithms such as the Gibbs sampler and the Metropolis–Hastings algorithm. It is interesting to see, in retrospect, that earlier papers such as Tanner and Wong (1987) and Besag and Clifford (1989) had proposed similar solutions (but did not receive the same response from the statistical community).

7.2 The two-stage Gibbs sampler

The *two-stage Gibbs sampler* creates a Markov chain from a joint distribution in the following way. If two random variables X and Y have joint density $f(x, y)$, with corresponding conditional densities $f_{Y|X}$ and $f_{X|Y}$, the two-stage Gibbs sampler generates a Markov chain (X_t, Y_t) according to the following steps:

Algorithm 7 Two-stage Gibbs sampler
Take $X_0 = x_0$
For $t = 1, 2, \ldots$, generate
    1. $Y_t \sim f_{Y|X}(\cdot|x_{t-1})$;
    2. $X_t \sim f_{X|Y}(\cdot|y_t)$.

Algorithm 7 is then straightforward to implement as long as simulating from both conditionals is feasible.[1] It is also easy to see why, if (X_t, Y_t) is distributed from f, then so is (X_{t+1}, Y_{t+1}), because both steps of iteration t use simulation from the true conditionals. Convergence of the Markov chain (and thus the algorithm) is therefore ensured unless the supports of the conditionals are not connected.

Example 7.1. To start with an obvious illustration, consider the bivariate normal model

$$(7.1) \qquad (X, Y) \sim \mathcal{N}_2 \left(0, \begin{pmatrix} 1 & \rho \\ \rho & 1 \end{pmatrix} \right),$$

for which the Gibbs sampler is
Given x_t, generate

$$Y_{t+1} \mid x_t \sim \mathcal{N}(\rho x_t, \ 1 - \rho^2),$$
$$X_{t+1} \mid y_{t+1} \sim \mathcal{N}(\rho y_{t+1}, \ 1 - \rho^2).$$

The subchain $(X_t)_t$ then satisfies

$$X_{t+1} | X_t = x_t \sim \mathcal{N}(\rho^2 x_t, \ 1 - \rho^4),$$

and a recursion shows that

$$X_t | X_0 = x_0 \sim \mathcal{N}(\rho^{2t} x_0, \ 1 - \rho^{4t}),$$

which does indeed converge to $\mathcal{N}(0, 1)$ as t goes to infinity.   ◀

As illustrated by the example above, the sequence (X_t, Y_t), $t = 1, \ldots, T$, produced by a Gibbs sampler converges to the *joint* distribution f and, as a consequence, both sequences $(X_t)_t$ and $(Y_t)_t$ converge to their respective *marginal distributions*.

Exercise 7.1 Show that the subsequence (X_t) resulting from Algorithm 7 is a Markov chain. (*Hint:* Use the fact that (X_t, Y_t) is generated conditional on X_{t-1} only.)

Perhaps the main reason why the Gibbs sampler became so popular in the 1990s as the reference MCMC algorithm is that it was the perfect computational complement to hierarchical models, which were then starting to be seriously investigated. As detailed and justified in Section 7.5, a hierarchical model specifies a joint distribution as successive layers of conditional distributions. The following example gives a first look at hierarchical models.

---

[1] When $f(x, y)$ is available in closed form, up to a normalizing constant, so are $f_{Y|X}$ and $f_{X|Y}$. Therefore, if simulating directly from those conditionals is not possible, Monte Carlo or MCMC approximations can be used, as developed in Section 7.6.3.

Example 7.2. Considering the pair of distributions

$$X|\theta \sim \mathcal{B}in(n,\theta), \quad \theta \sim \mathcal{B}e(a,b),$$

leads to the joint distribution

$$f(x,\theta) = \binom{n}{x} \frac{\Gamma(a+b)}{\Gamma(a)\Gamma(b)} \theta^{x+a-1}(1-\theta)^{n-x+b-1}.$$

The corresponding conditional distribution of $X|\theta$ is given above, while $\theta|x \sim \mathcal{B}e(x+a, n-x+b)$. The associated Gibbs sampler can be implemented as

```
> Nsim=5000                          #initial values
> n=15
> a=3
> b=7
> X=T=array(0,dim=c(Nsim,1))         #init arrays
> T[1]=rbeta(1,a,b)                  #init chains
> X[1]=rbinom(1,n,T[1])
> for (i in 2:Nsim){                 #sampling loop
+    X[i]=rbinom(1,n,T[i-1])
+    T[i]=rbeta(1,a+X[i],n-X[i]+b)
+    }
```

and its output is illustrated in Figure 7.1 for each marginal. Since this is a toy example, the closed-form marginals are available and thus produced on top of the histograms, and they show a good fit for both Gibbs samples.   ◄

Exercise 7.2 The marginal distribution of θ in Example 7.2 is the standard $\mathcal{B}e(a,b)$ distribution, but the marginal distribution of X is less standard and is known as the *beta-binomial* distribution.

a. Produce a closed-form expression for the beta-binomial density by integrating $f(x,\theta)$ in Example 7.2 with respect to θ.
b. Use this expression to create the function betabi in R. Then use the R command curve(betabi(x,a,b,n)) to draw a curve on top of the histogram as in Figure 7.1.

Example 7.3. Consider the posterior distribution on (θ, σ^2) associated with the joint model

(7.2)
$$X_i \sim \mathcal{N}(\theta, \sigma^2), \quad i = 1, \dots, n,$$
$$\theta \sim \mathcal{N}(\theta_0, \tau^2), \quad \sigma^2 \sim \mathcal{IG}(a,b),$$

where $\mathcal{IG}(a,b)$ is the inverted gamma distribution (that is, the distribution of the inverse of a gamma variable), with density $b^a(1/x)^{a+1}e^{-b/x}/\Gamma(a)$ and with

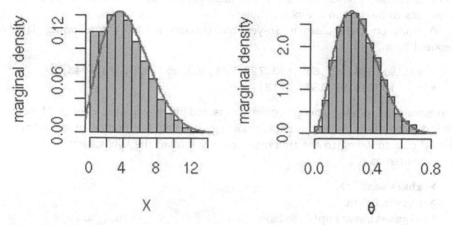

Fig. 7.1. Histograms of marginal distributions from the Gibbs sampler of Example 7.2 based on 5000 iterations of Algorithm 7 for $n = 15, a = 3, b = 7$. The true marginal distribution of θ is $\mathcal{B}e(a, b)$ and the marginal distribution of X is beta-binomial.

θ_0, τ^2, a, b specified. Writing $\mathbf{x} = (x_1, \ldots, x_n)$, the posterior distribution on (θ, σ^2) is given by

$$(7.3) \qquad f(\theta, \sigma^2 | \mathbf{x}) \propto \left[\frac{1}{(\sigma^2)^{n/2}} e^{-\sum_i (x_i - \theta)^2 / (2\sigma^2)} \right]$$
$$\times \left[\frac{1}{\tau} e^{-(\theta - \theta_0)^2 / (2\tau^2)} \right] \times \left[\frac{1}{(\sigma^2)^{a+1}} e^{1/b\sigma^2} \right],$$

from which we can get the full conditionals of θ and σ^2. (Note that this is not a regular conjugate setting in that integrating θ or σ^2 in this density does not produce a standard density.) Writing $\mathbf{x} = (x_1, \ldots, x_n)$, we have

$$\pi(\theta | \mathbf{x}, \sigma^2) \propto e^{-\sum_i (x_i - \theta)^2 / (2\sigma^2)} e^{-(\theta - \theta_0)^2 / (2\tau^2 \sigma^2)},$$

$$(7.4)$$

$$\pi(\sigma^2 | \mathbf{x}, \theta) \propto \left(\frac{1}{\sigma^2} \right)^{(n+2a+3)/2} e^{-\frac{1}{2\sigma^2} \left(\sum_i (x_i - \theta)^2 + (\theta - \theta_0)^2 / \tau^2 + 2/b \right)}.$$

These densities correspond to

$$\theta | \mathbf{x}, \sigma^2 \sim \mathcal{N} \left(\frac{\sigma^2}{\sigma^2 + n\tau^2} \theta_0 + \frac{n\tau^2}{\sigma^2 + n\tau^2} \bar{x}, \frac{\sigma^2 \tau^2}{\sigma^2 + n\tau^2} \right)$$

and

$$\sigma^2 | \mathbf{x}, \theta \sim \mathcal{IG} \left(\frac{n}{2} + a, \frac{1}{2} \sum_i (x_i - \theta)^2 + b \right),$$

where $\bar{x}$ is the empirical average of the observations, as the full conditional distributions to be used in a Gibbs sampler.

A study on metabolism in 15-year-old females yielded the following data, denoted by **x**,

```
> x=c(91,504,557,609,693,727,764,803,857,929,970,1043,
+      1089,1195,1384,1713)
```

corresponding to their energy intake, measured in megajoules, over a 24 hour period (also available in the dataset Energy). Using the normal model above, with θ corresponding to the true mean energy intake, the Gibbs sampler can be implemented as

```
> xbar=mean(x)
> sh1=(n/2)+a
> sigma=theta=rep(0,Nsim)                    #init arrays
> sigma{1}=1/rgamma(1,shape=a,rate=b)        #init chains
> B=sigma2{1}/(sigma2{1}+n*tau2)
> theta{1}=rnorm(1,m=B*theta0+(1-B)*xbar,sd=sqrt(tau2*B))
> for (i in 2:Nsim){
+    B=sigma2[i-1]/(sigma2[i-1]+n*tau2)
+    theta[i]=rnorm(1,m=B*theta0+(1-B)*xbar,sd=sqrt(tau2*B))
+    ra1=(1/2)*(sum((x-theta[i])^2))+b
+    sigma2[i]=1/rgamma(1,shape=sh1,rate=ra1)
+ }
```

where theta0, tau2, a, and b are specified values. The posterior means of θ and σ^2 are 872.402 and 136,229.2, giving as an estimate of σ 369.092. Histograms of the posterior distributions of $\log(\theta)$ and $\log(\sigma)$ are given in Figure 7.2.   ◀

Exercise 7.3 In connection with Example 7.3

a. Reproduce Figure 7.2 and superimpose the true marginal posteriors of $\log(\theta)$ and $\log(\sigma)$ by integrating $f(\theta, \sigma^2|x)$ in σ^2 and θ, respectively.
b. Investigate the sensitivity of the answer for a range of specifications of the hyperparameter values theta0, tau2, a, and b. Specifically, compute point estimates and confidence limits for θ and σ over a range of values for those parameters.

We want to point out that recognizing the full conditionals from a joint distribution is not that difficult. For example, the posterior distribution proportional to (7.3) is obtained by multiplying the densities in the specification (7.2).

To find a *full* conditional (that is, the conditional distribution of one parameter conditional on all others), we merely need to pick out all of the terms

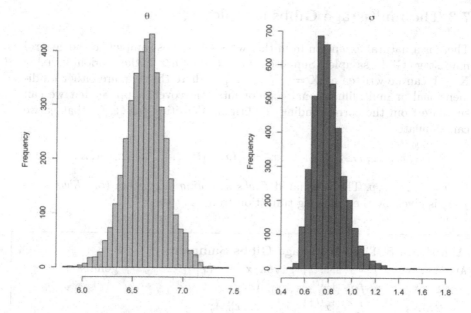

Fig. 7.2. Histograms of marginal posterior distributions of the log-mean and log-standard deviation from the Gibbs sampler of Example 7.3 based on 5000 iterations, with $a = b = 3$, $\tau_2 = 10$ and $\theta_0 = 5$. The 90% interval for $\log(\theta)$ is $(6.299, 6.960)$ and for $\log(\sigma)$ it is $(0.614, 1.029)$.

in the joint distribution that involve that parameter. For example, from (7.3), we see that

$$f(\theta|\sigma^2, \mathbf{x}) \propto \left[\frac{1}{(\sigma^2)^{n/2}} e^{-\sum_i (x_i - \theta)^2 / (2\sigma^2)} \right] \times \left[\frac{1}{\tau} e^{-(\theta - \theta_0)^2 / (2\tau^2)} \right],$$

$$f(\sigma^2|\theta, \mathbf{x}) \propto \left[\frac{1}{(\sigma^2)^{n/2}} e^{-\sum_i (x_i - \theta)^2 / (2\sigma^2)} \right] \times \left[\frac{1}{(\sigma^2)^{a+1}} e^{1/b\sigma^2} \right].$$

It should then be easy to see that the full conditional of σ^2 will be an inverted gamma distribution, as defined on page 202 (see also Exercise 7.19). For θ, although there is a little more algebra involved in the derivation, we can recognize that the full conditional will be normal. See Exercise 7.20 for an illustration with a larger hierarchy.

Exercise 7.4 Make explicit the derivations that connect the expressions above and the full conditional distributions in (7.4).

7.3 The multistage Gibbs sampler

There is a natural extension from the two-stage Gibbs sampler to the general multistage Gibbs sampler. Suppose that, for some $p > 1$, the random variable $\mathbf{X} \in \mathcal{X}$ can be written as $\mathbf{X} = (X_1, \ldots, X_p)$, where the X_i's are either unidimensional or multidimensional components. Moreover, suppose that we can simulate from the corresponding conditional densities $f_1, \ldots, f_p$, that is, we can simulate

$$X_i | x_1, x_2, \ldots, x_{i-1}, x_{i+1}, \ldots, x_p \sim f_i(x_i | x_1, x_2, \ldots, x_{i-1}, x_{i+1}, \ldots, x_p)$$

for $i = 1, 2, \ldots, p$. The associated *Gibbs sampling* algorithm (or *Gibbs sampler*) is given by the following transition from $X^{(t)}$ to $X^{(t+1)}$:

Algorithm 8 The Multistage Gibbs Sampler
At iteration $t = 1, 2, \ldots,$, given $\mathbf{x}^{(t)} = (x_1^{(t)}, \ldots, x_p^{(t)})$, generate
1. $X_1^{(t+1)} \sim f_1(x_1 | x_2^{(t)}, \ldots, x_p^{(t)})$;
2. $X_2^{(t+1)} \sim f_2(x_2 | x_1^{(t+1)}, x_3^{(t)}, \ldots, x_p^{(t)})$;
    $\vdots$
p. $X_p^{(t+1)} \sim f_p(x_p | x_1^{(t+1)}, \ldots, x_{p-1}^{(t+1)})$.

The densities $f_1, \ldots, f_p$ are called the *full conditionals*, and a particular feature of the Gibbs sampler is that these are the only densities used for simulation. Thus, even in a high-dimensional problem, *all of the simulations may be univariate*, which is usually an advantage.

Example 7.4. As an extension of Example 7.1, consider the multivariate normal density

$$(7.5) \qquad (X_1, X_2, \ldots, X_p) \sim \mathcal{N}_p\left(0, (1 - \rho)I + \rho J\right),$$

where I is the $p \times p$ identity matrix and J is a $p \times p$ matrix of ones. This is a model for *equicorrelation*, as $\mathrm{corr}(X_i, X_j) = \rho$ for every i and j. Using standard formulas for the conditional distributions of a multivariate normal random variable (see, for example, Johnson and Wichern, 1988), it is straightforward but tedious to verify that

$$X_i | x_{(-i)} \sim \mathcal{N}\left(\frac{(p-1)\rho}{1 + (p-2)\rho}\bar{x}_{(-i)}, \frac{1 + (p-2)\rho - (p-1)\rho^2}{1 + (p-2)\rho}\right),$$

where $x_{(-i)} = (x_1, x_2, \ldots, x_{i-1}, x_{i+1}, \ldots, x_p)$ and $\bar{x}_{(-i)}$ is the mean of this vector. The Gibbs sampler that generates from these univariate normals can then be easily derived, although it is useless for this problem (Exercise 7.5). It is, however, a short step to consider the setup where the components of the normal vector are restricted to a subset of $\mathbb{R}^p$. If this subset is a hypercube,

$$\mathfrak{H} = \prod_{i=1}^p (a_i, b_i),$$

then the corresponding conditionals simply are the normals above restricted to (a_i, b_i) for $i = 1, \ldots, p$ (in which case an exact algorithm such as sadmvn can be used). For more complex constraints, a Gibbs sampler is however (almost) required, as exact solutions do not exist. This Gibbs sampler is still based on normal full conditionals, which are now restricted to subsets of the real line and thus easily simulated (Exercise 2.22). ◀

Exercise 7.5 Given the normal target $\mathcal{N}_p(0, (1 - \rho)I + \rho J)$:

a. Write a Gibbs sampler using the conditional distributions provided in Example 7.4. Run your R code for $p = 5$ and $\rho = .25$, and verify graphically that the marginals are all $\mathcal{N}(0, 1)$.
b. Compare your algorithm using $T = 500$ iterations with rmnorm described in Section 2.2.1 in terms of execution time.
c. Propose a constrained subset that is not a hypercube, and derive the corresponding Gibbs sampler. (*Hint*: Consider, for example, a constraint such as $\sum_{i=1}^m x_i^2 \leq \sum_{i=m+1}^p x_i^2$ for $m \leq p - 1$.)

Models more complex than the one in Example 7.3 can be considered for the normal sampling model, as in the following case.

Example 7.5. A hierarchical specification for the normal model is the *one-way random effects model*. There are different ways to parameterize this model, but a possibility is as follows (see others in Example 7.14 and Exercise 7.24):

$$
\begin{aligned}
X_{ij} &\sim \mathcal{N}(\theta_i, \sigma^2), \quad i = 1, \ldots, k, \quad j = 1, \ldots, n_i, \\
\theta_i &\sim \mathcal{N}(\mu, \tau^2), \quad i = 1, \ldots, k, \\
\mu &\sim \mathcal{N}(\mu_0, \sigma_\mu^2), \\
\sigma^2 &\sim \mathcal{IG}(a_1, b_1), \quad \tau^2 \sim \mathcal{IG}(a_2, b_2), \quad \sigma_\mu^2 \sim \mathcal{IG}(a_3, b_3).
\end{aligned}
\tag{7.6}
$$

Now, if we proceed as before and write down the joint distribution from this hierarchy, we can derive the set of full conditionals

$$
\theta_i \sim \mathcal{N}\left(\frac{\sigma^2}{\sigma^2 + n_i \tau^2} \mu + \frac{n_i \tau^2}{\sigma^2 + n_i \tau^2} \bar{X}_i, \frac{\sigma^2 \tau^2}{\sigma^2 + n_i \tau^2} \right), \quad i = 1, \ldots, k,
$$

$$
\mu \sim \mathcal{N}\left(\frac{\tau^2}{\tau^2 + k \sigma_\mu^2} \mu_0 + \frac{k \sigma_\mu^2}{\tau^2 + k \sigma_\mu^2} \bar{\theta}, \frac{\sigma_\mu^2 \tau^2}{\tau^2 + k \sigma_\mu^2} \right),
$$

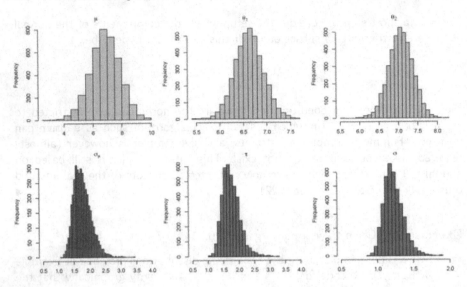

Fig. 7.3. Histograms of marginal posterior distributions from the Gibbs sampler of Example 7.5 based on 5000 iterations. The top row gives histograms for the underlying mean μ and the means, θ_1 and θ_2, for the girls' and boys' energy. The bottom row corresponds to the standard deviations.

$$(7.7) \quad \sigma^2 \sim \mathcal{IG}\left(n/2 + a_1, \ (1/2) \sum_{ij}(X_{ij} - \theta_i)^2 + b_1 \right),$$

$$\tau^2 \sim \mathcal{IG}\left(k/2 + a_2, \ (1/2) \sum_i(\theta_i - \mu)^2 + b_2 \right),$$

$$\sigma_\mu^2 \sim \mathcal{IG}\left(1/2 + a_3, \ (1/2)(\mu - \mu_0)^2 + b_3 \right),$$

where $n = \sum_i n_i$ and $\bar{\theta} = \sum_i n_i \theta_i / n$.

Expanding on the study in Example 7.3, the dataset Energy also contains data on the energy intake of boys. Model (7.6) applies (with $k = 2$) to the simultaneous analysis of the energy intakes of girls and boys. The outcome of the Gibbs sampler based on the conditionals in (7.7) is summarized in Figure 7.3. ◄

Exercise 7.6 In the setting of Example 7.5:

a. Derive the full conditional distributions in (7.7).
b. Implement this Gibbs sampler in R to reproduce the histograms in Figure 7.3.
c. A variation on the model (7.6) is to give μ a flat prior, which is equivalent to setting $\sigma_\mu^2 = \infty$ in (7.6). Construct the full conditionals for this model and modify the previous R code to compare both models on the Energy data.

7.4 Missing data and latent variables

Starting with the two-stage Gibbs sampler, working on a joint distribution $f(x, y)$, there seems to be a major difference with the Metropolis–Hastings algorithm that works with a single distribution or, in other words, generates all components of (x, y) at once. This difference in the target is illusory in that once given $f(x, y)$ we can use either the relevant Gibbs sampler or a generic Metropolis–Hastings algorithm, while if given a marginal density $f_X(x)$, we can construct (or *complete* $f_X(x)$) into a joint density $f(x, y)$ to aid in simulation, where the second variable Y is then an *auxiliary variable* that may not be directly relevant from a statistical point of view. There are many settings where a natural completion of $f_X(x)$ into $f(x, y)$ does exist and in fact this can lead to an effective Gibbs sampler.[2]

These considerations bring us back into the realm of *missing-data models*, as described in Section 5.4.2, where the representation (5.9)

$$g(x|\theta) = \int_Z f(x, z|\theta)\, dz$$

was introduced. As discussed in Chapter 5, $g(x|\theta)$ is the density of the observations (that is, the likelihood), and the right side represents the completion joint density. The density f is arbitrary and can be chosen so that the full conditionals of f are easy to simulate from and the Gibbs algorithm (Algorithm 8) is implemented on f instead of g and the corresponding full conditional of θ given (x, z).

Depending on the field, such representations go by different names. From a mathematical perspective, (5.9) is a mixture model. In statistics, we most often use the name of missing-data models, while econometricians prefer the use of *latent variable* models, maybe because of the related feeling of *deus ex machina* operating behind the scenes! If we factor $f(x, z|\theta) = f(x|z, \theta)h(z|\theta)$, then (5.9) becomes

$$g(x|\theta) = \int_Z f(x|z, \theta)h(z|\theta)\, dz,$$

and $h(z|\theta)$, the marginal distribution of the missing data z, is clearly a mixing distribution.

In a general missing-data setting,

$$g(x) = \int_Z f(x, z)\, dz$$

---

[2] It is obviously always the case that any given density $f_X(x)$ can be artificially completed into a joint density $f(x, y)$, as demonstrated with the slice sampler at the end of this section.

for $p \geq 2$, we write $y = (x, z) = (y_1, \ldots, y_p)$ and denote the conditional densities of $f(y) = f(y_1, \ldots, y_p)$ by

$$Y_1|y_2, \ldots, y_p \sim f_1(y_1|y_2, \ldots, y_p),$$
$$Y_2|y_1, y_3, \ldots, y_p \sim f_2(y_2|y_1, y_3, \ldots, y_p),$$
$$\vdots$$
$$Y_p|y_1, \ldots, y_{p-1} \sim f_p(y_p|y_1, \ldots, y_{p-1}).$$

Then, applying a multistage Gibbs sampler as in Algorithm 8 to those full conditionals and assuming they all can be simulated leads to a Markov $(Y^{(t)})_t$ that converges to f and therefore a subchain $(X^{(t)})_t$ that converges to g.

Example 7.6. In Examples 5.13 and 5.14, we treated a censored-data model as a missing-data model. We identify $g(x|\theta)$ with the likelihood function

$$g(x|\theta) = L(\theta|x) \propto \prod_{i=1}^{m} e^{-(x_i-\theta)^2/2},$$

and

$$f(x, z|\theta) = L(\theta|x, z) \propto \prod_{i=1}^{m} e^{-(x_i-\theta)^2/2} \prod_{i=m+1}^{n} e^{-(z_i-\theta)^2/2}$$

is the complete-data likelihood. Given a prior distribution $\pi(\theta)$ on θ, we can then create a Gibbs sampler that iterates between the conditional distributions

$$\pi(\theta|x, z) \quad \text{and} \quad f(z|x, \theta)$$

and will have stationary distribution $\pi(\theta, z|x)$, the posterior distribution of (θ, z).
 Taking a flat prior $\pi(\theta) = 1$, the conditional distribution of $\theta|x, z$ is given by

$$\theta|x, z \sim \mathcal{N}\left(\frac{m\bar{x} + (n - m)\bar{z}}{n}, \frac{1}{n}\right),$$

while the conditional distribution of $Z|x, \theta$ is the product of the truncated normals

$$Z_i|x, \theta \sim \varphi(z - \theta) \Big/ \{1 - \Phi(a - \theta)\},$$

as each Z_i must be greater than the truncation point a. Generating values of Z can be done via the R function rtrun from the package bayesm (see Exercises 7.21 and 7.7). The outcome of the Gibbs sampler, whose R core can be written as

```
> for(i in 2:Nsim){
>   zbar[i]=mean(rtrun(mean=rep(that[i-1],n-m),
+   sigma=rep(1,n-m),a=rep(a,n-m),b=rep(Inf,n-m)))
>   that[i]=rnorm(1,(m/n)*xbar+(1-m/n)*zbar[i],sqrt(1/n))
>   }
```

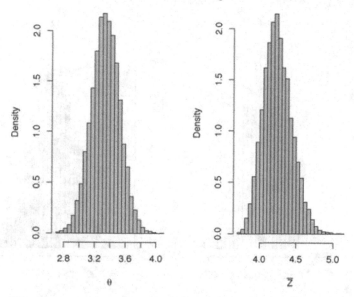

Fig. 7.4. Histograms of the posterior distributions of θ and $\bar{Z}$ from Example 7.6. The truncation point on the Z_i's is $a = 3.5$.

is summarized in Figure 7.4 using the posterior distributions of θ and $\bar{Z}$.   ◀

Exercise 7.7 Referring to Example 7.6:

a. Show that, as a function of θ, the complete data likelihood is proportional to the density of $\mathcal{N}(\{m\bar{x} + (n-m)\bar{z}\}/n, 1/n)$.
b. Complete the R code above into a Gibbs sampler that estimates the posterior distribution of θ.

Example 7.7. Recall the multinomial model of Example 5.16,

$$\mathcal{M}\left(n; \frac{1}{2} + \frac{\theta}{4}, \frac{1}{4}(1-\theta), \frac{1}{4}(1-\theta), \frac{\theta}{4}\right).$$

where we estimated θ using either EM or MCEM steps, introducing the latent variable Z with the demarginalization

$$(z, x_1 - z, x_2, x_3, x_4) \sim \mathcal{M}\left(n; \frac{1}{2}, \frac{\theta}{4}, \frac{1}{4}(1-\theta), \frac{1}{4}(1-\theta), \frac{\theta}{4}\right).$$

If we use a uniform prior on θ, the full conditionals can be recovered as

$$\theta \sim \mathcal{B}e(z + x_4 + 1, x_2 + x_3 + 1) \text{ and } z \sim \mathcal{B}\text{in}\left(x_1, \frac{\theta}{2+\theta}\right),$$

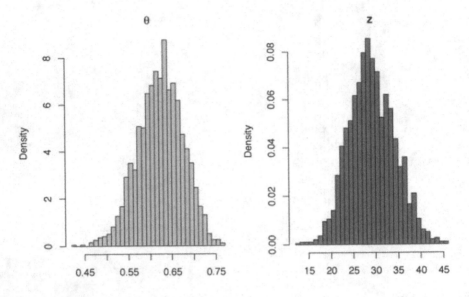

Fig. 7.5. Histograms of marginal distributions from the Gibbs sampler of Example 7.7. The main interest is in the marginal distribution of θ.

leading to the Gibbs sampler

```
> x=c(125,18,20,34)              #data
> theta=z=rep(.5,Nsim)          #init chain
> for (j in 2:Nsim){
>    theta[j]=rbeta(1,z[j-1]+x[4]+1,x[2]+x[3]+1)
>    z[j]=rbinom(1,x{1},(theta[j]/(2+theta[j])))
>    }
```

whose output is summarized in Figure 7.5.   ◄

This example shows a case where both EM and the Gibbs sampler apply. As usual, the Bayesian approach allows a more complete inference that includes confidence intervals.

Exercise 7.8 In the setting of Example 7.7:

a. Construct a 95% confidence interval for θ based on the outcome of the Gibbs sampler, and verify whether or not the EM solution belongs to this interval.
b. The Gibbs sampler above used a uniform prior on θ; that is, $\theta \sim Be(a,b)$ with $a = b = 1$. Write a Gibbs sampler for general a and b, and, for a range of a and b, compare the Gibbs estimates of θ with the EM answer. What can you conclude about sensitivity to the prior?

Example 7.8. A generalization of the model of Example 7.7 is the model

$$(7.8) \quad X \sim \mathcal{M}_5\left(n; a_1\theta_1 + b_1, a_2\theta_1 + b_2, a_3\theta_2 + b_3, a_4\theta_2 + b_4, c(1 - \theta_1 - \theta_2)\right),$$

with $0 \le a_1 + a_2 = a_3 + a_4 = 1 - \sum_{i=1}^{4} b_i = c \le 1$, where the $a_i, b_i \ge 0$ are known based on genetic considerations, as in Table 7.1 describing the probabilities of the four blood types as functions of genotype probabilities, because of allele dominance. Our interest is in estimating the allele frequencies p_A, p_B, and p_O (which sum to 1).

We can then augment the data with $\mathbf{Z} = (Z_1, Z_2, Z_3, Z_4)$ as

$$X_1 = Z_1 + Z_2, \quad X_2 = Z_3 + Z_4, \quad X_3 = Z_5 + Z_6, \quad X_4 = Z_7 + Z_8,$$

which demarginalizes the model to allow us to sample from

$$Y \sim \mathcal{M}_9\left(n; a_1\theta_1, b_1, a_2\theta_1, b_2, a_3\theta_2, b_3, a_4\theta_2, b_4, c(1 - \theta_1 - \theta_2)\right),$$

with $Y = (Z_1, X_1 - Z_1, Z_2, X_2 - Z_2, Z_3, X_3 - Z_3, Z_4, X_4 - Z_4, X_5)$. (See Exercise 7.23 for an alternate solution.) A natural prior distribution on (θ_1, θ_2) is the Dirichlet prior $\mathcal{D}(\alpha_1, \alpha_2, \alpha_3)$,

$$\pi(\theta_1, \theta_2) \propto \theta_1^{\alpha_1 - 1} \theta_2^{\alpha_2 - 1} (1 - \theta_1 - \theta_2)^{\alpha_3 - 1},$$

which leads to the full conditionals

$$(7.9) \quad \begin{aligned} (\theta_1, \theta_2, 1 - \theta_1 - \theta_2)|x, \mathbf{z} &\sim \mathcal{D}(z_1 + z_2 + \alpha_1, z_3 + z_4 + \alpha_2, x_5 + \alpha_3), \\ Z_i|x, \theta_1, \theta_2 &\sim \mathcal{B}\left(x_i, \frac{a_i\theta_1}{a_i\theta_1 + b_i}\right) \quad (i = 1, 3), \\ Z_i|x, \theta_1, \theta_2 &\sim \mathcal{B}\left(x_i, \frac{a_i\theta_2}{a_i\theta_2 + b_i}\right) \quad (i = 5, 7), \end{aligned}$$

which can all easily be simulated and thus included within a Gibbs sampler. Figure 7.6 shows the distributions of chains produced by such a sampler.   ◀

Table 7.1.   Observed genotype frequencies on blood type data. The effect of a dominant allele creates a missing-data problem.

Genotype	Probability	Observed	Probability	Frequency
AA	p_A^2	A	$p_A^2 + 2p_A p_O$	$n_A = 186$
AO	$2p_A p_O$			
BB	p_B^2	B	$p_B^2 + 2p_B p_O$	$n_B = 38$
BO	$2p_B p_O$			
AB	$2p_A p_B$	AB	$p_A p_B$	$n_{AB} = 13$
OO	p_O^2	O	p_O^2	$n_O = 284$

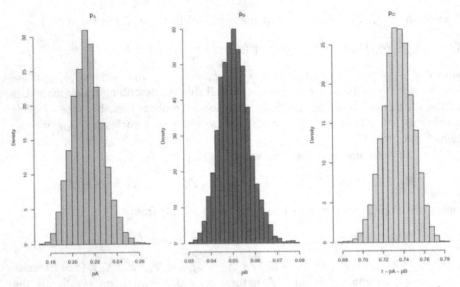

Fig. 7.6. Histograms of marginal distributions of the genotype probabilities from the Gibbs sampler of Example 7.8.

Exercise 7.9 For the data in Table 7.1, modeled with (7.8):

a. Verify that the observed data likelihood is proportional to

$$(p_A^2 + 2p_A p_O)^{n_A} (p_B^2 + 2p_B p_O)^{n_B} (p_A p_B)^{n_{AB}} (p_O^2)^{n_O}.$$

b. With missing data Z_A and Z_B, verify that the complete-data likelihood is proportional to

$$(p_A^2)^{Z_A} (2p_A p_O)^{n_A - Z_A} (p_B^2)^{Z_B} (2p_B p_O)^{n_B - Z_B} (p_A p_B)^{n_{AB}} (p_O^2)^{n_O}.$$

c. Write a Gibbs sampler to estimate p_A and p_B.

Finite mixture models, which we have seen in some detail in Chapter 5 (Example 5.2) and Chapter 6 (Example 6.5), are obviously a candidate for demarginalization through latent variables; that is, as a special case of a mixture (!). As already described in Example 5.12, given a sample $(x_1, \ldots, x_n)$ from a mixture distribution

$$\sum_{j=1}^{k} p_j \, f(x|\xi_j) \,,$$

where $\sum_j p_j = 1$ and $f(\cdot|\xi_j)$ is a parameterized density with unknown parameter ξ_j, we can associate with every observation x_i a latent variable

$z_i \in \{1, \ldots, k\}$ that indicates which component of the mixture is associated with x_i. The corresponding completion of the mixture model above is then

$$Z_i \sim \mathcal{M}_k(1; p_1, \ldots, p_k), \qquad x_i | z_i \sim f(x|\xi_{z_i}).$$

Thus, considering $y_i = (x_i, z_i)$ (instead of x_i) entirely eliminates the mixture structure since the likelihood of the completed model is

$$\ell(p, \xi | y_1, \ldots, y_n) \propto \prod_{i=1}^{n} p_{z_i}\, f(x_i | \xi_{z_i}) = \prod_{j=1}^{k} \prod_{i; z_i = j} p_j\, f(x_i | \xi_j).$$

One may wonder why the completion is useful in this setting since the observed likelihood can be computed in closed form, as shown for instance in Figure 5.2, which represents a mixture likelihood on a grid of pixels as in Example 6.5, where we produced a random walk Metropolis–Hastings algorithm. As in the EM algorithm of Examples 5.12 and 5.13, using the latent indicator variables produces a usually efficient simulation algorithm that quickly focuses on the mode(s) of the posterior distribution.

The two steps of the Gibbs sampler are then associated with the full conditional posteriors

$$P(Z_i = j | \mathbf{x}, \xi) \propto p_j\, f(x_i | \xi_j) \qquad (i = 1, \ldots, n, \; j = 1, \ldots, k)$$

and

$$\xi_j | \mathbf{y} \sim \pi\left(\xi \middle| \frac{\lambda_j \alpha_j + n_j \bar{x}_j}{\lambda_j + n_j}, \lambda_j + n_j\right),$$
$$p \sim \mathcal{D}_k(\gamma_1 + n_1, \ldots, \gamma_k + n_k),$$

where

$$n_j = \sum_{i=1}^{n} \mathbb{I}_{z_i = j}, \qquad n_j \bar{x}_j = \sum_{i=1}^{n} \mathbb{I}_{z_i = j} x_i.$$

In this two-step Gibbs sampler, the generation from the posterior associated with the complete likelihood is not detailed, as it will vary depending on the sampling model and the prior used. In the standard situation relying on an exponential family for $f(\cdot|\xi)$ and a conjugate prior on ξ, this generation is obviously straightforward.

Example 7.9. As an illustration, consider the same setting as in Example 5.12, namely a normal mixture with two components with equal known variance and fixed weights,

$$p \mathcal{N}(\mu_1, \sigma^2) + (1 - p) \mathcal{N}(\mu_2, \sigma^2).$$

We assume in addition a normal $\mathcal{N}(0, v^2 \sigma^2)$ prior distribution, with v^2 known, on both means μ_1 and μ_2. The latent variables z_i are the same as in Example 5.12, namely

$$P(Z_i = 1) = 1 - P(Z_i = 2) = p \quad \text{and} \quad X_i|Z_i = k \sim \mathcal{N}(\mu_k, \sigma^2).$$

The completed distribution is then

$$\pi(\mu_1, \mu_2, \mathbf{z}|\mathbf{x}) \propto \exp\left\{-(\mu_1^2 + \mu_2^2)/v^2\sigma^2\right\}$$
$$\times \prod_{i:z_i=1} p \exp\left\{-\frac{(x_i - \mu_1)^2}{2\sigma^2}\right\} \prod_{i:z_i=2} (1-p) \exp\left\{-\frac{(x_i - \mu_2)^2}{2\sigma^2}\right\},$$

for which the full conditionals of the μ_j's are easily derived (Exercise 7.10). Figure 7.7 illustrates the behavior of the corresponding Gibbs sampler using a simulated dataset $\mathbf{x}$ of 500 points from the $.7\mathcal{N}(0,1)+.3\mathcal{N}(2.7,1)$ distribution. This picture plots the MCMC sample after $15,000$ iterations on top of the log-posterior surface. This simulation is in fact in clear agreement with the posterior surface. Although it may appear to be too concentrated around one mode, you must account for the fact that the second mode represented on this graph is much lower since there is a difference of at least 50 in log-posterior values.   ◄

Exercise 7.10 Using the completed joint distribution in Example 7.9:

a. Show that the conditional distributions are $j = 1, 2$

$$\mu_j|\mathbf{x}, \mathbf{z} \sim \mathcal{N}\left(\frac{v^2}{n_j v^2 + 1}\sum_{i;z_i=j} x_i, \ \frac{\sigma^2 v^2}{n_j v^2 + 1}\right),$$

where n_j denotes the number of z_i's equal to j and

$$P(Z_i = j|x_i, \mu_1, \mu_2) = \frac{p\exp\left\{-\frac{(x_i-\mu_j)^2}{2\sigma^2}\right\}}{p\exp\left\{-\frac{(x_i-\mu_1)^2}{2\sigma^2}\right\} + (1-p)\left\{-\frac{(x_i-\mu_2)^2}{2\sigma^2}\right\}}.$$

b. Write the R code to reproduce Figure 7.7.
c. For $\sigma = 1$, investigate the convergence of the Gibbs sampler for various combinations of the true values of (μ_1, μ_2, p). In particular, you should find that if the μ_is are too separated and $p = 0.5$, the Gibbs sampler may concentrate in one mode even though the modal likelihoods are similar.

As a last "example" of a latent variable Gibbs sampler, we look at the *slice sampler*, which appears more like a generic type of demarginalization.[3] (See Neal, 2003, for a comprehensive treatment.) Given a density of interest $f_X(x)$, we can always represent it as the marginal density of the joint density

---

[3] In fact, we can alternatively consider the Gibbs sampler as being derived from the slice sampler; see Robert and Casella (2004, Chapter 8).

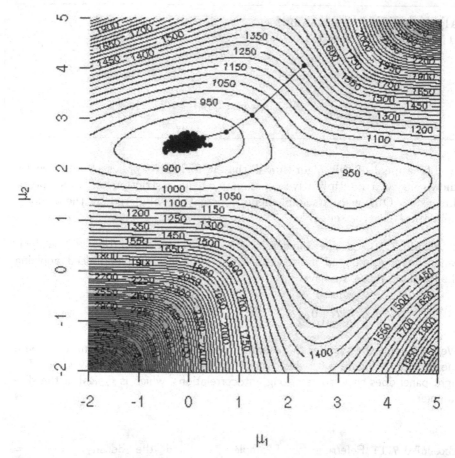

Fig. 7.7. Gibbs sample of 5000 points for the mixture posterior against the log-posterior surface.

$$f(x, u) = \mathbb{I}\{0 < u < f_X(x)\}$$

since integrating the above in u returns f_X. The associated conditional densities are

$$f_{X|U}(x|u) = \frac{\mathbb{I}\{0 < u < f_X(x)\}}{\int \mathbb{I}\{0 < u < f_X(x)\}\, dx}, f_{U|X}(u|x) = \frac{\mathbb{I}\{0 < u < f_X(x)\}}{\int \mathbb{I}\{0 < u < f_X(x)\}\, du},$$

which means they are both uniform. Those two conditionals then define the slice sampler as the associated Gibbs sampler.

Algorithm 9 2D slice sampler
At iteration t, simulate

1. $U^{(t+1)} \sim \mathcal{U}_{[0, f(x^{(t)})]}$;
2. $X^{(t+1)} \sim \mathcal{U}_{A^{(t+1)}}$, with

$$A^{(t+1)} = \{x : f(x) \geq u^{(t+1)}\}.$$

The appeal of this algorithm is that it formally applies to any density known up to a multiplicative constant with no restriction on its shape or dimension. Obviously, its implementation may be hindered by the uniform simulation over the set $A^{(t)}$.

Example 7.10. Consider the density $f(x) = \frac{1}{2}e^{-\sqrt{x}}$ defined for $x > 0$. While it can be directly simulated, it also yields easily to the slice sampler. Indeed, applying the formulas above, we have

$$U|x \sim \mathcal{U}\left(0, \frac{1}{2}e^{-\sqrt{x}}\right), \quad X|u \sim \mathcal{U}\left(0, [\log(2u)]^2\right).$$

We implement the sampler to generate 5000 variates and plot them along with the density in Figure 7.8, which shows that the agreement is very good. The right panel does show some strong autocorrelations, which is typical of the slice sampler.   ◀

Exercise 7.11 Referring to Example 7.10 and the density $f_X(x) = (1/2)\exp(-\sqrt{x})$:

a. Verify that the conditional distributions are

$$U|x \sim \mathcal{U}\left(0, (1/2)\exp(-\sqrt{x})\right) \text{ and } X|u \sim \mathcal{U}\left(0, [\log(2u)]^2\right),$$

and implement a Gibbs sampler to generate random variables from $f_X(x)$.
b. Make the transformation $Y = \sqrt{X}$ and show that $Y \sim \mathcal{G}(3/2, 1)$. Use this fact to simulate directly X. Compare this algorithm with the slice sampler.

There is an obvious extension to the 2D slice sampler above, akin to the multistage extension to the two-stage Gibbs sampler. If the target density is written as a product of functions,

$$f(x) = \prod_{i=1}^{n} g_i(x),$$

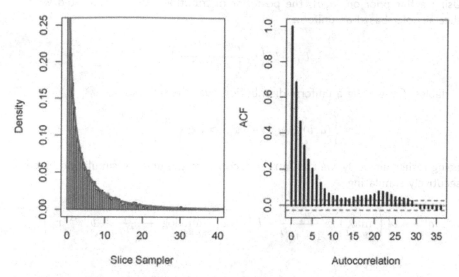

Fig. 7.8. A slice sampler histogram and target density for Example 7.10 using 5000 iterations. The left panel is the histogram with the true density overlaid, and the right panel shows the autocorrelation function.

as for instance in the case of a posterior distribution associated with a sample of n observations (where the g_i's are then the componentwise densities), an associated completion is

$$f(x, u_1, \ldots, u_n) = \prod_{i=1}^{n} \mathbb{I}\{0 < u_i < g_i(x)\},$$

which leads to a slice sampler with $(n+1)$ steps, $X^{(t)}$ then being uniformly generated over the set

$$A^{(t)} = \bigcap_{i=1}^{n} \left\{ x : g_i(x) > u_i^{(t)} \right\}.$$

Example 7.11. Recall logistic regression, which we first saw in Example 4.11 and fit with a Metropolis–Hastings algorithm in Exercise 6.13. The model is

$$Y_i \sim \text{Bernoulli}(p(x_i)), \quad p(x) = \frac{\exp(\alpha + \beta x)}{1 + \exp(\alpha + \beta x)},$$

where $p(x)$ is the success probability and x is a unidimensional covariate. The likelihood associated with a sample $(\mathbf{y}, \mathbf{x}) = (y_1, x_1), \ldots, (y_n, x_n)$ is

$$L(\alpha, \beta | \mathbf{y}) \propto \prod_{i=1}^{n} \left(\frac{e^{\alpha + \beta x_i}}{1 + e^{\alpha + \beta x_i}} \right)^{y_i} \left(\frac{1}{1 + e^{\alpha + \beta x_i}} \right)^{1-y_i}.$$

Using a flat prior on (a, b), the posterior distribution can be associated with a slice sampler based on uniform

$$U_i \sim \mathcal{U}\left(0, \frac{e^{y_i(\alpha+\beta x_i)}}{1 + e^{\alpha+\beta x_i}}\right)$$

variables. Generating a uniform distribution over the set

$$\left\{(a, b) : y_i(a + bx_i) > \log \frac{u_i}{1 - u_i}\right\}$$

being rather unwieldy, we can further decompose the uniform simulation by consecutively simulating

$$a^{(t)} \sim \mathcal{U}\left(\max_{i;y_i=1} \log \frac{u_i^{(t)}}{1 - u_i^{(t)}} - b^{(t-1)}x_i, \min_{i;y_i=0} \log \frac{1 - u_i^{(t)}}{u_i^{(t)}} - b^{(t-1)}x_i\right)$$

and

$$b^{(t)} \sim \mathcal{U}\left(\max_{i;y_i=1}\left[\log \frac{u_i^{(t)}}{1 - u_i^{(t)}} - a^{(t)}\right]/x_i, \min_{i;y_i=0}\left[\log \frac{1 - u_i^{(t)}}{u_i^{(t)}} - a^{(t)}\right]/x_i\right),$$

if we assume without loss of generality that all x_i's are positive. However, running the corresponding slice sampler on the challenger dataset described in Exercise 6.13 exhibits a random walk behavior on the chain $(a^{(t)}, b^{(t)})_t$, as shown in Figure 7.9. We therefore introduce instead normal $\mathcal{N}(0, \sigma^2)$ priors on both a and b. The modification on the slice sampler is minimal in that both uniform distributions above are replaced with truncated normals $\mathcal{N}(0, \sigma^2)$, the truncation intervals being those used above. The core of the R code is then

```
> for (t in 2:Nsim){
+    uni=runif(n)*exp(y*(a[t-1]+b[t-1]*x))/
+              (1+exp(a[t-1]+b[t-1]*x))
+    mina=max(log(uni[y==1]/(1-uni[y==1]))-b[t-1]*x[y==1])
+    maxa=min(-log(uni[y==0]/(1-uni[y==0]))-b[t-1]*x[y==0])
+    a[t]=rtrun(0,sigmaa,mina,maxa)
+    minb=max((log(uni[y==1]/(1-uni[y==1]))-a[t])/x[y==1])
+    maxb=min((-log(uni[y==0]/(1-uni[y==0]))-a[t])/x[y==0])
+    b[t]=rtrun(0,sigmab,minb,maxb)
+    }
```

with sigmaa equal to 5 and sigmab equal to 5 divided by the standard deviation of the x_i's.   ◀

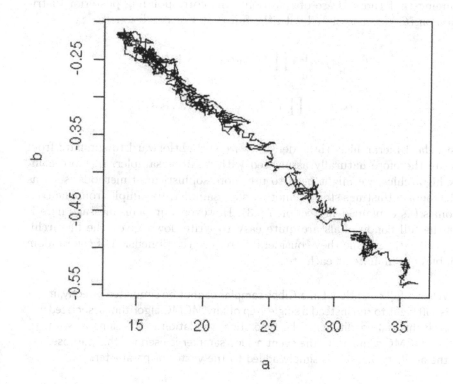

Fig. 7.9. Evolution of the chain $(a^{(t)}, b^{(t)})_t$ along 10^3 final iterations of a slice sampler for the **challenger** dataset under a flat prior.

7.5 Hierarchical structures

We have seen the multistage Gibbs sampler applied to a number of examples, most arising from missing-data structures. However, it is equally well-suited to sample in a straightforward way from any hierarchical model.

A *hierarchical model* is defined by a sequence of conditional distributions as, for instance, in the two-level generic hierarchy

$$X_i \sim f_i(x|\theta), \quad i = 1, \ldots, n, \quad \theta = (\theta_1, \ldots, \theta_p),$$
$$\theta_j \sim \pi_j(\theta|\gamma), \quad j = 1, \ldots, p, \quad \gamma = (\gamma_1, \ldots, \gamma_s),$$
$$\gamma_k \sim g(\gamma), \quad k = 1, \ldots, s.$$

The joint distribution from this hierarchy is

$$\prod_{i=1}^{n} f_i(x_i|\theta) \prod_{j=1}^{p} \pi_j(\theta_j|\gamma) \prod_{k=1}^{s} g(\gamma_k).$$

Assuming that the x_i's are observations, the corresponding posterior distribution on (θ, γ) is associated with the full posterior conditionals

$$\theta_j \propto \pi_j(\theta_j|\gamma) \prod_{i=1}^{n} f_i(x_i|\theta), \quad j = 1, \ldots, p,$$

$$\gamma_k \propto g(\gamma_k) \prod_{j=1}^{p} \pi_j(\theta_j|\gamma), \quad k = 1, \ldots, s.$$

In standard hierarchies, these densities are straightforward to simulate from and are therefore naturally associated with a Gibbs sampler. In more complex hierarchies, we might need to use more sophisticated methods, such as a Metropolis–Hastings step or another slice sampler, to sample from the conditionals (as explained in Section 7.6.3). However, our main message here is that the full conditionals are quite easy to write down given the hierarchical specification, while they considerably reduce the dimension of the random variables to simulate at each step.

⚡ When a full conditional in a Gibbs sampler cannot be simulated directly, it is sufficient to run instead a single step of any MCMC algorithm associated with this full conditional. The theoretical validation is the same as with any MCMC sampler. In the event a slice sampler is used for this purpose, the auxiliary variable is simply added to the vector of parameters.

Example 7.12. A benchmark hierarchical example in the Gibbs sampling literature describes multiple failures of ten pumps in a nuclear plant, with the data given in Table 7.2. The modeling is based on the assumption that the number of

Table 7.2. Number of failures and times of observation of ten pumps in a nuclear plant (*source*: Gaver and O'Muircheartaigh, 1987).

Pump	1	2	3	4	5	6	7	8	9	10
Failures	5	1	5	14	3	19	1	1	4	22
Time	94.32	15.72	62.88	125.76	5.24	31.44	1.05	1.05	2.10	10.48

failures of the ith pump follows a Poisson process with parameter λ_i ($1 \leq i \leq 10$). For an observation time t_i, the number of failures X_i is thus a Poisson $\mathcal{P}(\lambda_i t_i)$ random variable. The standard prior distributions are gamma distributions, which lead to the hierarchical model

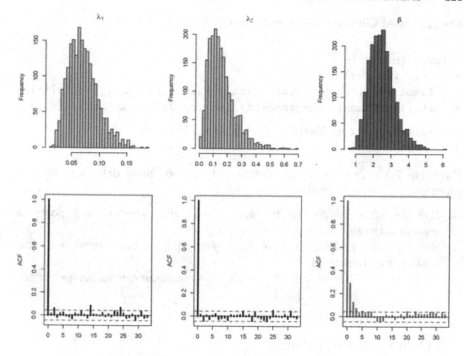

Fig. 7.10. Histograms of marginal distributions of λ_1, λ_2, and β from the pump failure data of Example 7.12. The corresponding bottom panels are autocorrelation plots. The hyperparameter values are $\alpha = 1.8$, $\gamma = 0.01$, and $\delta = 1$.

$$X_i \sim \mathcal{P}(\lambda_i t_i), \quad i = 1, \ldots 10,$$
$$\lambda_i \sim \mathcal{G}(\alpha, \beta), \quad i = 1, \ldots 10,$$
$$\beta \sim \mathcal{G}(\gamma, \delta).$$

The joint distribution is thus

$$\pi(\lambda_1, \ldots, \lambda_{10}, \beta | t_1, \ldots, t_{10}, p_1, \ldots, p_{10})$$

$$\propto \prod_{i=1}^{10} \left\{ (\lambda_i t_i)^{x_i} \, e^{-\lambda_i t_i} \, \lambda_i^{\alpha-1} e^{-\beta \lambda_i} \right\} \beta^{10\alpha} \beta^{\gamma-1} e^{-\delta\beta}$$

$$\propto \prod_{i=1}^{10} \left\{ \lambda_i^{x_i + \alpha - 1} \, e^{-(t_i + \beta)\lambda_i} \right\} \beta^{10\alpha + \gamma - 1} e^{-\delta\beta},$$

leading to the full conditional distributions

$$\lambda_i | \beta, t_i, x_i \sim \mathcal{G}(x_i + \alpha, t_i + \beta), \quad i = 1, \ldots 10,$$

$$\beta | \lambda_1, \ldots, \lambda_{10} \sim \mathcal{G}\left(\gamma + 10\alpha, \delta + \sum_{i=1}^{10} \lambda_i \right).$$

The associated Gibbs sampler is quite straightforward, with core R code

```
> for(i in 2:Nsim){
+   for(j in 1:nx)
+     lambda[i,j]=rgamma(1,sh=xdata[j]+alpha,ra=Time[j]+beta[i-1])
+   beta[i]=rgamma(1,sh=gamma+nx*alpha,ra=delta+sum(lambda[i,]))}
```

The result of a run over 5000 iterations is shown in Figure 7.10.   ◄

Exercise 7.12 One reason for collecting the pump failure data is to identify which pumps are more reliable.

a. Run the Gibbs sampler for the pump failure data and get 95% posterior credible intervals for the parameters λ_i.
b. Based on the analysis, can you identify any pumps that are more or less reliable than the others?
c. How does your answer in b. change as the hyperparameter values are varied?

7.6 Other considerations

In this last section, we look at a few issues that could arise in the implementation of a Gibbs sampler.

7.6.1 Reparameterization

Many factors contribute to the convergence properties of a Gibbs sampler. For example, convergence performance may be greatly affected by the choice of the coordinates (or, in other words, the parameterization). If the covariance matrix Σ of the target has a wide range of eigenvalues, the Gibbs sampler may be very slow to explore the entire range of the support of the target.

Example 7.13. Recall Example 7.1, where we saw a first Gibbs sampler for the bivariate normal in (7.1). For that bivariate normal distribution, Figure 7.11 shows the autocorrelation for $\rho = .3, .6, .9$. The higher correlation results in a sampler that will have more trouble exploring the entire space and thus require more iterations. It is also interesting to note that no matter what is the value of ρ, $X + Y$ and $X - Y$ are independent, and thus changing coordinates from (x, y) to $(x + y, x - y)$ would lead to an immediately converging Gibbs algorithm.   ◄

Exercise 7.13 For the bivariate normal distribution (7.1):

a. prove that $X + Y$ and $X - Y$ are independent.

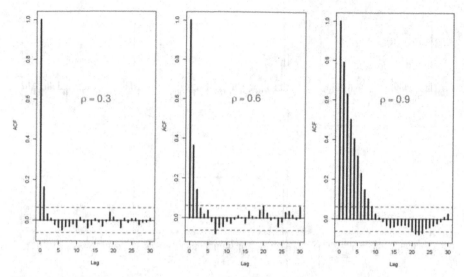

Fig. 7.11. Autocorrelations in one marginal of a bivariate normal generated from a Gibbs sampler for $\rho = 0.3$ (left), $\rho = 0.6$ (middle), and $\rho = 0.9$ (right).

b. Suppose now that X and Y are bivariate normal with mean 0, correlation ρ, and $\text{var}(X) = \sigma_x^2$ and $\text{var}(T) = \sigma_y^2$, which are not necessarily equal. Study the effect on autocorrelation of varying ρ, σ_x^2, and σ_y^2.

c. If $\sigma_x^2 \neq \sigma_y^2$, then $X + Y$ and $X - Y$ are no longer independent. Find a pair of random variables that are.

Convergence of both Gibbs sampling and Metropolis–Hastings algorithms may thus suffer from a poor choice of parameterization. As a result of this, the MCMC literature has considered changes in the parameterization of a model as a way to speed up convergence in a Gibbs sampler. It seems, however, that most efforts have concentrated on the improvement of specific models, resulting in a lack of general methodology for the choice of a "proper" parameterization. Nevertheless, the overall advice is to try to make the components "as independent as possible" and to use several parameterizations simultaneously to intermingle the conditionals.

Example 7.14. (Continuation of Example 7.5) A reparameterization of the one-way random effect of Example 7.5 is to introduce the overall mean at the observation level, as in

$$
\begin{aligned}
X_{ij} &\sim \mathcal{N}(\mu + \theta_i, \sigma^2), \quad i = 1, \ldots, k, \quad j = 1, \ldots, n_i, \\
\theta_i &\sim \mathcal{N}(0, \tau^2), \quad i = 1, \ldots, k, \\
\mu &\sim \mathcal{N}(\mu_0, \sigma_\mu^2).
\end{aligned}
$$

(7.10)

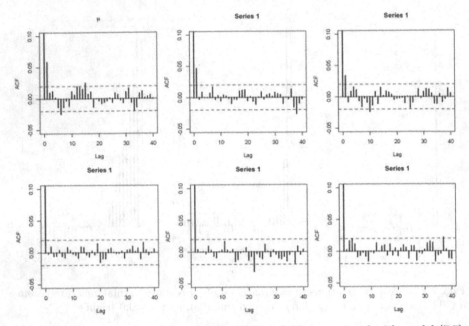

Fig. 7.12. Autocovariance plots for the Gibbs sampler associated with model (7.7) and the Gibbs sampler associated with its reparameterization (7.10). The top row gives the autocovariances for μ, θ_1, θ_2 (left to right) for model (7.7), and the bottom row gives them for model (7.10).

While the hierarchy looks the same, the conditionals are different (Exercise 7.14) and the properties of the corresponding Gibbs sampler are as well. When applied to the Energy dataset, the new Gibbs sampler is not as good. For instance, Figure 7.12 shows the autocorrelations, which, to the eye, seem only slightly better for the first model. However, if we look at the covariance matrix of the subchain $(\mu^{(t)}, \theta_1^{(t)}, \theta_2^{(t)})$, its estimate is

$$
\begin{pmatrix} 1.056 & -0.175 & -0.166 \\ -0.175 & 1.029 & 0.018 \\ -0.166 & 0.018 & 1.026 \end{pmatrix} \text{ and } \begin{pmatrix} 1.604 & 0.681 & 0.698 \\ 0.681 & 1.289 & 0.278 \\ 0.698 & 0.278 & 1.304 \end{pmatrix},
$$

for model (7.7) and model (7.10), respectively, so the variances and covariances are larger for the reparameterized model. Thus, we clearly should use the parameterization of model (7.7).   ◀

Exercise 7.14 For the reparameterized model of (7.10):

a. Show that the full conditionals of θ_i and μ are

$$\theta_i \sim \mathcal{N}\left(B_1(\bar{X}_i - \mu), (\sigma^2/n_i)B_1\right), \quad B_1 = \frac{n_i\tau^2}{n_i\tau^2 + \sigma^2}, \quad i = 1, \ldots, k$$

$$\mu \sim \mathcal{N}\left((1 - B_2)\mu_0 + B_2(\bar{X} - \bar{\theta}), (\sigma^2/n)B_2\right), \quad B_2 = \frac{n\sigma_\mu^2}{n\sigma_\mu^2 + \sigma^2},$$

where $n = \sum_i n_i$ and $\bar{\theta} = \sum_i n_i\theta_i/n$.

b. Write a Gibbs sampler for this model, and compare the autocovariances with those of the Gibbs sampler based on model (7.7).

c. The covariance matrix of the parameter estimates is the inverse of the Fisher information matrix. Calculate this matrix for both parameterizations using the R functions cor and solve.

7.6.2 Rao–Blackwellization

We have already seen Rao–Blackwellization in Section 4.6, where conditioning on a subset of the simulated variables may produce considerable improvement upon the standard empirical estimator in terms of variance by a simple "recycling" of the rejected variables. However, as the Gibbs sampler accepts every simulated value, this type of recycling cannot apply. Nonetheless, Gelfand and Smith (1990) propose a type of conditioning that we will call *parametric Rao–Blackwellization* to differentiate it from the form studied in Section 4.6.

For $(X, Y) \sim f(x, y)$, *parametric Rao–Blackwellization* is based on the marginalization identity (iterated expectation)

$$\mathbb{E}[X] = \mathbb{E}[\mathbb{E}[X|Y]].$$

Defining $\delta(Y) = \mathbb{E}[X|Y]$, we have $\mathbb{E}[\delta(Y)] = \mathbb{E}[X]$ and $\text{var}[\delta(Y)] \leq \text{var}(X)$, showing that $\delta(Y)$ is the better estimator (provided it can be computed).

Example 7.15. (Continuation of Example 7.1) In the case where the target is the bivariate normal distribution, the full conditionals are

$$X_{t+1} \mid y_t \sim \mathcal{N}(\rho y_t, \ 1 - \rho^2),$$
$$Y_{t+1} \mid x_{t+1} \sim \mathcal{N}(\rho x_{t+1}, \ 1 - \rho^2),$$

and thus it follows that $\mathbb{E}[X|Y] = \rho Y$. Since X and Y have the same marginal distribution, the variance of the Rao–Blackwellized version is then obviously reduced by a factor ρ^2.   ◀

Unfortunately, the variance reduction from using δ_Y does not hold in general, due to the correlation in the MCMC sample. However, Liu et al. (1994) have shown that, in particular, the improvement holds for any two-stage Gibbs sampler.

We now look at another example of Rao–Blackwellization in a missing-data Gibbs sampler for a common occurrence where possible gains occur.

Example 7.16. For 360 consecutive time units, consider recording the number of passages of individuals per unit time past some sensor. This can be, for instance, the number of cars observed at a crossroad. Hypothetical results are

Number of passages	0	1	2	3	4 or more
Number of observations	139	128	55	25	13

The data involves a grouping of the observations with four passages or more. This can be addressed as a missing-data model, where we assume that the ungrouped observations are $X_i \sim \mathcal{P}(\lambda)$. The likelihood of the model is

$$\ell(\lambda|x_1,\ldots,x_5) \propto e^{-347\lambda}\lambda^{128+55\times2+25\times3}\left(1 - e^{-\lambda}\sum_{i=0}^{3}\lambda^i/i!\right)^{13}$$

for $x_1 = 139,\ldots,x_5 = 13$. For $\pi(\lambda) = 1/\lambda$ and $\mathbf{z} = (z_1,\ldots,z_{13})$, the vector of the 13 units larger than 4, we can derive a completion Gibbs sampler from the full conditionals

$$Z_i^{(t)} \sim \mathcal{P}(\lambda^{(t-1)})\,\mathbb{I}_{y\geq4}, \quad i = 1,\ldots,13,$$

$$\lambda^{(t)} \sim \mathcal{G}\left(313 + \sum_{i=1}^{13} Z_i^{(t)},\ 360\right).$$

The Rao–Blackwellized estimate of λ is then given by

$$\sum_{t=1}^{T} \mathbb{E}\left[\lambda\big|,z_1^{(t)},\ldots,z_{13}^{(t)}\right] = \frac{1}{360T}\sum_{t=1}^{T}\left(313 + \sum_{i=1}^{13} y_i^{(t)}\right),$$

and the evolution of this estimator, along with the empirical average, is shown in Figure 7.13. It exhibits a massive variance reduction.   ◄

Exercise 7.15 Referring to Example 7.16:

a. Verify the likelihood function and the Gibbs sampler.
b. Write R code to reproduce Figure 7.13.
c. The truncated Poisson variable can be generated using the while statement

```
> for (i in 1:13){while(y[i]<4) y[i]=rpois(1,lam[j-1])}
```

or directly with

```
> prob=dpois(c(4:top),lam[j-1])
> for (i in 1:13) z[i]=4+sum(prob<runif(1)*sum(prob))
```

Compare the efficiencies of these two algorithms. In theory, the value of top should be infinity. In practice, what value would you use?

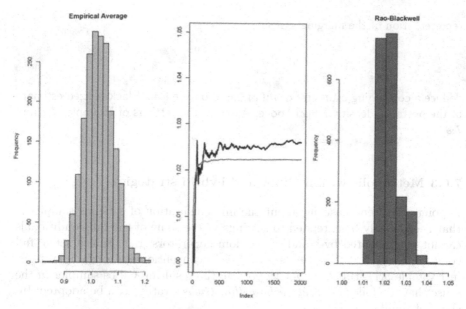

Fig. 7.13. For the counting data of Example 7.16, the histogram of λ (left) and its conditional expectation $\mathbb{E}(\lambda|Z) = 313 + \sum_{i=1}^{13} z_i$ (right). Note the difference in scale of the histograms. The center panel shows the evolution of the cumulative averages of the empirical average (black) and the Rao–Blackwellization (grey).

There also exist non-parametric Rao–Blackwellized estimators in missing-variable settings. When considering an approximation to the marginal distribution f_X associated with $f(x, y_1, \ldots, y_p)$, a Rao–Blackwellized estimator associated with the Gibbs chain $(x^{(t)}, \mathbf{y}^{(t)})_t$ is given by

$$(7.11) \qquad \widehat{f}_X(x) = \frac{1}{T} \sum_{t=1}^{T} f(x|\mathbf{y}^{(t)}),$$

which converges at a parametric speed to $f_X(x)$. This estimator gives a smooth approximation to the marginal, which can be plotted on top of the marginal.

As studied in Exercises 3.15, 4.1, and 4.2, the approximation of the Bayes factors calls for specific solutions. Chib (1995) proposes an alternative approach based on a Rao–Blackwellization that is much more efficient when it can be implemented.

Exercise 7.16 In a missing-variable setting where the sampling density can be written as

$$f(x|\theta) = \int_{\mathcal{Z}} g(x, z|\theta) \, dz,$$

we assume the prior $\pi(\theta)$ is such that a two-stage Gibbs sampler based on the simulation of $g(z|x, \theta)$ and $\pi(\theta|x, z)$ can be implemented. Using a Bayes' Theorem representation of the marginal density,

$$m(x) = \frac{f(x|\theta)\pi(\theta)}{\pi(\theta|x)},$$

deduce a converging estimator of $m(x)$ based on the Rao–Blackwellized estimate of the posterior density $\pi(\theta|x)$ above. Apply to the settings of Examples 7.7 and 7.9.

7.6.3 Metropolis within Gibbs and hybrid strategies

A point worth emphasizing about the implementation of a Gibbs sampler is that it can easily be extended to settings where some of the full conditionals cannot be simulated by standard random generators. If, within a set of full conditionals $f_1, \ldots, f_p$, some density f_i is unconventional, for example (5.15) in Example 5.17, this does not jeopardize the resulting Gibbs sampler in the sense that the following *Metropolis-within-Gibbs* strategy can be adopted: Instead of simulating

$$X_i^{(t+1)} \sim f_i(x_i | x_1^{(t+1)}, \ldots, x_{i-1}^{(t+1)}, x_{i+1}^{(t)}, \ldots, x_p^{(t)}),$$

you can run *one single step* of any MCMC scheme associated with the stationary distribution $f_i(x_i | x_1^{(t+1)}, \ldots, x_{i-1}^{(t+1)}, x_{i+1}^{(t)}, \ldots, x_p^{(t)})$. A simple solution is for instance to use a random walk Metropolis algorithm centered at $x_i^{(t)}$. While at first this sounds like a crude approximation, as the full conditional is not *exactly* simulated, the validity of the resulting algorithm is exactly the same as with the original Gibbs sampler since the joint distribution f remains the stationary distribution of the corresponding Markov chain.

You may then wonder what the point is in using a Gibbs sampler if componentwise simulations have to be replaced with Metropolis–Hastings steps, as using a Metropolis–Hastings algorithm targeted at the joint distribution f is more "natural". While there is nothing restraining you from using a joint Metropolis–Hastings algorithm, it is most often the case that designing such a Metropolis–Hastings algorithm on a large-dimensional target is challenging or even impossible. The fundamental gain in using a Gibbs-like structure is that it breaks down a complex model into a large number of smaller *and* simpler targets, where local Metropolis–Hastings algorithms can be designed at little expense.

Example 7.17. If we consider the target distribution (5.15), we mentioned in Example 5.17 that this is not a standard distribution. While Booth and Hobert (1999) designed a specific Accept-Reject algorithm to simulate from (5.15), a random walk proposal on each u_i, as in

```
> for (i in 1:n){
+   mu=u[i]
+   u[i]=factor*sigma[iter-1]*rnorm(1)+mu
+   if (log(runif(1))>gu(u[i],i,beta[iter-1],sigma[iter-1])-
+                       gu(mu,i,beta[iter-1],sigma[iter-1])){
+        u[i]=mu
+        }
+   }
```

produces a sample of u_i's at iteration iter conditional on the current values of the parameters and the sample of u_i's at iteration iter-1. In the overall Gibbs sampler, the parameters are then simulated by

```
> sigma=c(sigma,1/sqrt(2*rgamma(1,0.5*n)/sum(u^2)))
> tau=sigma[iter]/sqrt(sum(as.vector(x^2)*pro(beta[iter-1],u)))
> betaprop=beta[iter-1]+rnorm(1)*factor*tau
> if (log(runif(1))>likecomp(betaprop,sigma[iter],u)-
+                     likecomp(beta[iter-1],sigma[iter],u))
+       betarop=beta[iter-1]
> beta=c(beta,betaprop)
```

in a straightforward manner. (See Example 8.1 for the complete implementation.) The calibration term factor can further be tuned against the acceptance rate of Section 6.5, as described in Section 8.5.   ◄

   While remaining close to this idea of incorporating Metropolis–Hastings steps when direct simulation is not possible, we may also signal the possible extension to *hybrid strategies*.[4] The concept is once again based on the stationarity of the right target distribution, even though intuition may disagree. When given a (univariate or multivariate) target where several natural MCMC schemes are available, a hybrid algorithm merges those different schemes altogether. Schematically, if local or global (meaning componentwise or joint) MCMC update functions mcmc.1(x,y), ..., mcmc.q(x,y) are available, the transition kernel defined by

```
mcmc(x,y)=function(x,y){
  switch(sample(1:p,1),
  mcmc.1(x,y)
      ...
  mcmc.p(x,y))
  }
```

remains a valid MCMC update function against the *same* target distribution. While this sounds like a ludicrous idea because poor schemes are mixed

---

[4] Hybrid strategies should not be confused with *hybrid Monte Carlo* (Neal, 1999), also called Hamiltonian MCMC, which is a form of Langevin implementation aimed at reducing the waste of simulation in random walk proposals.

with good ones, the blind mixing of all available strategies is nonetheless (a) valid from the perspective of producing the correct stationary distribution and (b) risk-free in the sense that if the list of functions contains a single well-performing algorithm, the hybrid version will perform at least as well, simply requiring a p-fold extension of the computing time. For instance, if several blocking or reparameterization strategies are simultaneously available, they can all be incorporated within the same algorithm. This solution could well appear as a waste of computing time, but our advice on this matter is that, unless some of the mcmc.i functions clearly do work, the time spent (wasted) running the hybrid solution is time saved on designing and selecting the more efficient mcmc.i functions. In other words, it is more efficient to let the computer sort among the available solutions than to run preliminary tests to sort those solutions "by hand".

7.6.4 Improper priors

This section discusses a particular danger resulting from careless use of the Gibbs sampler. We know that the Gibbs sampler is based on conditional distributions derived from the joint distribution. However, what is particularly insidious is that these conditional distributions may be well-defined and may be simulated from but may not correspond to any joint distribution!

This problem is not a *defect* of the Gibbs sampler, or even a simulation problem, but rather a problem of inadvertently using the Gibbs sampler in a situation for which the underlying assumptions are violated. It is nonetheless important to warn the user of MCMC algorithms against this danger because it corresponds to a situation often encountered in Bayesian noninformative (or *"default"*) models.

The construction of the Gibbs sampler directly from the conditional distributions is a strong incentive to bypass checking for the propriety of the posterior, especially in complex setups. But such checking is essential, as the following simple example shows.

Example 7.18. The following model was used by Casella and George (1992) to point out the difficulty of assessing the impropriety of a posterior distribution through the conditional distributions. The pair of conditional densities

$$(7.12) \qquad X|y \sim \mathcal{E}xp(y), \qquad Y|x \sim \mathcal{E}xp(x),$$

are well-defined conditional distributions, but these conditional distributions do not correspond to any joint probability distribution. Figure 7.14 shows a histogram and cumulative average for a sample generated using the Gibbs sampler corresponding to those conditionals. The pictures are extremely curious and in fact are absolute rubbish! (This is not a recurrent Markov chain.) Indeed, the only function that could be the joint distribution is

$$f(x, y) \propto \exp(-xy),$$

which does not have a finite integral.   ◄

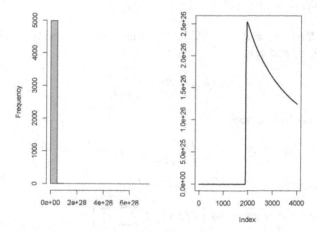

Fig. 7.14. Histogram and cumulative average of the X variable from the Gibbs sampler of (7.12). Note the ranges on the graphs that signal convergence issues.

Exercise 7.17 For the Gibbs sampler based on (7.12)

a. Write an R program to reproduce Figure 7.14.
b. The Hammersley–Clifford Theorem (Robert and Casella, 2004, Section 9.1.4) says that the joint density must satisfy

$$f(x, y) = f(y|x) \Big/ \int [f(y|x)/f(x|y)] \, dy .$$

Show that applying this result to (7.12) leads to $f(x, y) \propto \exp(-xy)$.
c. Show that if the exponential distributions are restricted to $(0, B)$, $B < \infty$, the resulting figure is reasonable. Exhibit the stationary density of the Markov chain in this case. (*Hint:* Apply the Hammersley–Clifford Theorem.)

Given the results of Example 7.18, it may appear that a simple graphical monitoring is enough to exhibit deviant behavior of the Gibbs sampler. However, this is not the case in general and there are many examples, some of which are published (see Casella, 1996), where the output of the Gibbs sampler seemingly does not differ from a convergent Markov chain. Often, this phenomenon takes place when the divergence of the posterior density occurs "at 0"; that is, at a specific point whose immediate neighborhood is rarely visited by the chain, as in the following random effects example. The only way to make sure the Gibbs sampler you are using is valid is to check that the joint distribution has a finite integral.

Example 7.19. Consider a random effects model,

$$Y_{ij} = \beta + U_i + \varepsilon_{ij}, \qquad\qquad i = 1, \ldots, I, \; j = 1, \ldots, J,$$

where $U_i \sim \mathcal{N}(0, \sigma^2)$ and $\varepsilon_{ij} \sim \mathcal{N}(0, \tau^2)$. The Jeffreys (improper) prior for the parameters β, σ, and τ is

$$\pi(\beta, \sigma^2, \tau^2) = \frac{1}{\sigma^2 \tau^2}.$$

The conditional distributions

$$U_i | y, \beta, \sigma^2, \tau^2 \sim \mathcal{N}\left(\frac{J(\bar{y}_i - \beta)}{J + \tau^2 \sigma^{-2}}, (J\tau^{-2} + \sigma^{-2})^{-1} \right),$$

$$\beta | u, y, \sigma^2, \tau^2 \sim \mathcal{N}(\bar{y} - \bar{u}, \tau^2 / JI),$$

$$\sigma^2 | u, \beta, y, \tau^2 \sim \mathcal{IG}\left(I/2, (1/2) \sum_i u_i^2 \right),$$

$$\tau^2 | u, \beta, y, \sigma^2 \sim \mathcal{IG}\left(IJ/2, (1/2) \sum_{i,j} (y_{ij} - u_i - \beta)^2 \right),$$

are well-defined, and a Gibbs sampler can be easily implemented in this setting. However, there is no proper joint distribution that corresponds to these conditionals! And, in many instances, as you may check for yourself, this is impossible to detect by monitoring the output.   ◄

Exercise 7.18 In the setting of Example 7.19:

a. Generate data according to the model and run a corresponding Gibbs sampler on the parameters of the model. Monitor histograms and cumulative averages. Can you detect the fact that there is no proper joint distribution?
b. The variation on the model (7.6) given in Exercise 7.6, where μ is given a flat prior, is a Gibbs sampler with improper priors. Since there is no guarantee that the posterior distribution is proper, check to see if it is in fact proper.

⚡ If improper priors are used in a Gibbs sampler, the posterior must *always* be checked for propriety. However, it is often the case that improper priors on variances cause more trouble than those on means.

7.7 Additional exercises

Exercise 7.19 The gamma distribution with parameters a and b, $\mathcal{G}(a, b)$, has density $b^a x^{a-1} e^{-bx} / \Gamma(a)$. Show that if $X \sim \mathcal{IG}(a, b)$, then $1/X \sim \mathcal{G}(a, b)$. (This means that generating from a gamma distribution is equivalent to generating from an inverted gamma distribution.)

Exercise 7.20 From the hierarchy (7.6), show that the joint distribution can be obtained by multiplying the densities together. Then, using the strategy of Exercise 7.4, verify that the full conditionals are given by (7.7).

Exercise 7.21 A truncated normal generator is based on the R function

```
rtnorm=function(n=1,mu=0,lo=-Inf,up=Inf){
  qnorm(runif(n,min=pnorm(lo,mean=mu,sd=sigma),
    max=pnorm(up,mean=mu,sd=sigma)),
    mean=mu,sd=sigma)}
```

where mu and sigma are the mean and standard deviation of the normal, lo is the lower truncation point, up is the upper truncation point, and n is the number of random variables desired. For $Z \sim \mathcal{N}(0, 1)$ with truncation (*i*) $-1 < Z < 1$, (*ii*) $Z < 1$, and (*iii*) $Z > 3$, generate 1000 random variables and compare the histograms with the density functions.

Exercise 7.22 Referring to Exercise 7.5:

a. Calculate the third and fourth moments of the density in question a of that exercise.
b. If $\mathbf{X} \sim \mathcal{N}_p(0, \Sigma)$, show that the density of $X_1|x_{(-1)}$ is

$$\mathcal{N}_p(\Sigma_{12}\Sigma_{22}^{-1}x_{(-1)}, \Sigma_{11} - \Sigma_{12}\Sigma_{22}^{-1}\Sigma_{12}'),$$

where the covariance matrix is partitioned in the obvious way. Use this formula to verify (7.5).
c. The matrix $(1 - \rho)I + \rho J$ is only positive definite if $\rho > -1/(p - 2)$. Verify this result.

Exercise 7.23 Referring to model (7.8), the (uncompleted) posterior distribution is available as

$$\pi(\theta_1, \theta_2|x) \propto (a_1\theta_2 + b_1)^{x_1}(a_2\theta_2 + b_2)^{x_2}(a_3\theta_1 + b_3)^{x_3}(a_4\theta_1 + b_4)^{x_4}$$
$$\times (1 - \theta_1 - \theta_2)^{x_5+\alpha_3-1}\theta_2^{\alpha_1-1}\theta_1^{\alpha_2-1}.$$

a. Show that the marginal distributions $\pi(\theta_1|x)$ and $\pi(\theta_2|x)$ can be explicitly computed as polynomials when the α_i's are integers.
b. Give the marginal posterior distribution of $\xi = \theta_2/(1-\theta_1-\theta_2)$. (*Note:* See Robert, 1995a, for a solution.)
c. Evaluate the Gibbs sampler based on (7.9) by comparing approximate moments of $\theta_1, \theta_1,$ and ξ with their exact counterparts derived from the explicit marginal.

Exercise 7.24 The alternate parameterization of model (7.6) produced in Example 7.14 modifies the relations between the variables. Show that θ_i and μ are a priori independent for this parameterization and that this is not the case in model (7.6).

Exercise 7.25 Rao–Blackwellization can be applied to most of the Gibbs samplers in this chapter. For each of the following examples, verify the conditional expectations provided there and compare via an R experiment the empirical average with the Rao–Blackwellization.

a. Example 7.2: $\mathbb{E}[\theta|x] = x + a/(n + a + b)$.

b. Equation (7.4): $\mathbb{E}[\theta|\mathbf{x}, \sigma^2] = \frac{\sigma^2}{\sigma^2 + n\tau^2}\, \theta_0 + \frac{n\tau^2}{\sigma^2 + n\tau^2}\, \bar{x}$.

c. Equation (7.7): $\mathbb{E}[\theta_i|\bar{X}_i, \sigma^2] = \frac{\sigma^2}{\sigma^2 + n_i\tau^2}\mu + \frac{n_i\tau^2}{\sigma^2 + n_i\tau^2}\bar{X}_i$.

d. Example 7.6: $\mathbb{E}[\theta|x, z] = \frac{m\bar{x} + (n-m)\bar{z}}{n}$.

e. Example 7.12: $\mathbb{E}[\lambda_i|\beta, t_i, x_i] = (x_i + \alpha)/(t_i + \beta)$.

8

Convergence Monitoring and Adaptation for MCMC Algorithms

"Why does he insist that we must have a diagnosis? Some things are not meant to be known by man."

Susanna Gregory
An Unholy Alliance

Reader's guide

The goal of this chapter is to present different monitoring methods (or *diagnostics*) proposed to check (for) the convergence of an MCMC algorithm when considering its output and to answer the most commonly asked question about MCMC, namely "when do we stop our MCMC algorithm?" We distinguish here between two separate notions of convergence, namely convergence to stationarity and convergence of ergodic average, in contrast with iid settings. We also discuss several types of convergence diagnostics, primarily those contained in the coda package of Plummer et al. (2006), even though more accurate methods may be available in specific settings.

Since assessing convergence is a preliminary step in comparing algorithms when several are considered, we also cover in this final chapter (albeit briefly) adaptive MCMC algorithms, pointing out the dangers of basic adaptivity and discussing the amcmc package developed by Rosenthal (2007) for a specific adaptive random walk proposal.

C.P. Robert, G. Casella, *Introducing Monte Carlo Methods with R*, Use R,
DOI 10.1007/978-1-4419-1576-4_8, © Springer Science+Business Media, LLC 2010

8.1 Introduction

In each of the previous two chapters, we have introduced the MCMC algorithms and, although we skipped most theoretical details, we argued that, under fairly general conditions, these algorithms are convergent because the chains they produce are ergodic. While such developments are obviously necessary as a theoretical validation of the MCMC algorithms, they are nonetheless insufficient from the point of view of the implementation of MCMC methods. Indeed, theoretical guarantees do not tell us when to stop these algorithms and when to produce our estimates with enough confidence. The ideal setting would be to provide you with clear convergence markers that could be included in your R program so that no sequential processing would be needed! In practice, this is nearly impossible, and several runs of your program are usually required until you are satisfied with the outcome (or you run out of time and/or patience).

As in Chapter 4, the techniques we present in this chapter are mostly empirical, with additional difficulties because we are only approximately simulating from the target density. We are mostly in the situation of describing a sequence of incomparable techniques with widely varying degrees of theoretical justification and usefulness, while the assessment derived from those techniques is partly subjective and clearly not foolproof.

Historically, there was a flurry of papers at the end of the 1990s concerned with the development of convergence diagnoses (see Robert and Casella, 2004, Chapter 13), including the construction of an R package called coda, which we will discuss below. This flurry has now quieted down, the main reason sadly being that no criterion is absolute. It is somewhat of an illusion to think we can control the flow of a Markov chain and assess its convergence behavior based only on a few (hundred or thousand or even million) realizations of this chain. There always are settings that, for most realizations, will invalidate an arbitrary diagnostic, and the randomness inherent in the nature of the problem prevents any categorical guarantee of performance.

8.2 Monitoring what and why

Before describing the monitoring tools, we need to discuss the target of those tools, that is, the three different types of convergence for which assessment is necessary.

8.2.1 Convergence to the stationary distribution

The first requirement for the convergence of an MCMC algorithm is that the distribution of the chain $(x^{(t)})$ should be the stationary distribution f. This sounds like a minimal requirement for an algorithm that was introduced as an approximate generator from f! Unfortunately, this issue is not as straightforward as it seems. Indeed, assessing that, for a given t, $x^{(t)}$ is exactly distributed from f is first a difficult goal when f is a complex target and second a theoretical impossibility if we only consider a *single realization* (or *path*) of the Markov chain $(x^{(t)})$. Both these issues can be settled by aiming at a slightly less ambitious goal, namely to experimentally assess the independence from the starting point $x^{(0)}$ based on several realizations of the chain using the same transition kernel.

In Robert and Casella (2004), we assumed that, in a standard statistical setup where the support of f is approximately known, this primary convergence issue is marginal, namely that the initial value $x^{(0)}$ can be chosen (as if) generated from the distribution f. When looking at a single chain, this means acting as if the chain is already in its stationary regime at the start, meaning in practice that $x^{(0)}$ belongs to an area of likely (enough) values for f. This assumption is harder to maintain in higher dimensions, however, when the support of f is generally unknown.

⚡ When running an MCMC algorithm, the important issues are the speed of exploration of the support of f (that is, the region that supports most of the probability mass of f) and the degree of correlation between the $x^{(t)}$'s. However, this is not to say that stationarity should not be tested. As Section 8.3.2 will explain, regardless of the starting distribution, the chain may be slow to explore the different regions of the support of f, with lengthy stays in some subregions. A stationarity test comparing several chains may then be useful in detecting such difficulties.

The major tool for assessing (convergence to) stationarity is to use several chains in parallel in order to compare their performances. Obviously, this means that the slower chain in the group governs the convergence diagnostic and that the choice of the initial distribution is extremely important in guaranteeing that the different chains are well-dispersed (with regard to the target distribution f). But the availability of multiprocessor machines is an incentive for using parallel chains since the Markovian nature of MCMC algorithms prevents the use of parallel processing. Checking for the convergence of an MCMC algorithm by using several chains at once is therefore not much more costly than using a single chain, while the output of all the chains can be recycled later for approximation (Monte Carlo) purposes.

Looking at a single path of the Markov chain produced by an MCMC algorithm makes it difficult to assess convergence unless something is known about the target f or the transition kernel K. In general, MCMC algorithms suffer from the major defect that *"you've only seen where you've been"* in the sense that the part of the support of f that has not (yet) been visited by the chain at time T is almost impossible to detect. This "missing mass" problem is a quite central difficulty with most MCMC algorithms, much more than the failure to explore the tails of f.

8.2.2 Convergence of averages

Once the issue of convergence to the stationary distribution is (approximately) settled, we are faced with the same goal as in regular Monte Carlo settings, namely convergence of the empirical average

$$(8.1) \qquad \frac{1}{T} \sum_{t=1}^{T} h(x^{(t)})$$

to $\mathbb{E}_f[h(X)]$ for an arbitrary function h. While the tools derived in Chapter 4 can be considered in this setting, there are two features that distinguish stationary MCMC outcomes from iid ones, namely the probabilistic dependence in the sample and, consequently, the mixing behavior of the transition, that is, how fast the chain explores the support of f.

While exploring the support of f may sound more like a stationarity constraint, we insist on separating these issues. An ergodic Markov chain that starts from one mode of f can be correctly deemed to be in its stationary regime, but it may still require many iterations to reach the other modes of this distribution, another occurrence of the "missing mass" problem. Thus, ergodicity, understood in the sense of an independence from initial conditions, is not to be confused with ergodicity taken in the sense of correct approximation of f by the empirical distribution.

Additional assessments are thus necessary to guarantee that the standard diagnostics of Chapter 4 are still appropriate in the Markovian case. For instance, the strong reliance on the Central Limit Theorem found in Chapter 4 needs to be tempered by the fact that the Central Limit Theorem does not hold for all ergodic Markov chains (Meyn and Tweedie, 1993). This second convergence perspective is therefore the one central to the assessment and comparison of MCMC algorithms.

8.2.3 Approximating iid sampling

Ideally, the approximation to f provided by MCMC algorithms should extend to the (approximate) production of iid samples from f. The most practical solution to this independence issue is to use *subsampling* (or *batch sampling*) to reduce or eliminate correlation between the successive points of the Markov chain. This technique subsamples the chain $(x^{(t)})$ with a (deterministic or random) batch size k, considering only the values $y^{(t)} = x^{(kt)}$. If the covariance $\mathrm{cov}_f(x^{(0)}, x^{(t)})$ decreases monotonically with t, the motivation for subsampling is obvious. However, checking for the monotone decrease of $\mathrm{cov}_f(x^{(0)}, x^{(t)})$ is not always possible and, in some settings, the covariance oscillates with t, which complicates the choice of k.

As noted in Chapter 4, assessing convergence almost inevitably leads to a "waste" of simulation in that it takes longer to diagnose convergence than it takes for convergence to "occur". Subsampling illustrates this general feature since it loses in efficiency with regard to the second convergence goal.

Exercise 8.1 Suppose both $\mathbb{E}_f[h(X)]$ and $\mathbb{E}_f[h^2(X)]$ are well-defined and $(x^{(t)})$ is a Markov chain with stationary distribution f. We compare an estimator that uses all of the Markov chain (δ_1) with one that subsamples (δ_2):

$$\delta_1 = \frac{1}{Tk} \sum_{t=1}^{Tk} h(x^{(t)}) \quad \text{and} \quad \delta_k = \frac{1}{T} \sum_{\ell=1}^{T} h(x^{(k\ell)}).$$

Show that the variance of δ_1 satisfies $\mathrm{var}(\delta_1) \leq \mathrm{var}(\delta_k)$ for every $k > 1$.

In the remainder of the chapter, we consider only independence issues in cases where they have bearing on diagnoses.

8.2.4 The coda package

Following the review in Cowles and Carlin (1996) and Brooks and Roberts (1998) of major convergence diagnostics, Plummer et al. (2006) have written an R package called coda that contains many of the tools we will be discussing in this chapter. While this coda package was primarily intended for processing the output of a BUGS run (Lunn et al., 2000), it can also be utilized directly to handle an arbitrary output from your own MCMC programs. Given that we will use some coda functions in the following sections, we need to introduce the basic features of this package at this early stage, referring you to the latest version of the coda manual for further details. (The manual is available in the same CRAN depository as the package.)

The coda package is available for download from the CRAN depository and works on all platforms.[1] Once installed on your machine, you can download all coda functions using the command library(coda).

The generic functions of the coda package are:

codamenu	gives a menu interface to the coda functions
coda.options	queries and sets options of the coda functions
multimenu	gives additional plotting options
read.coda	reads data in the format provided by BUGS

---

[1] Since the package is open-source, it is part of the major Linux distributions such as Ubuntu, Red Hat, and Debian.

and an essential family of functions called mcmc. These functions transform
an MCMC output made of a vector or a matrix into an MCMC object that
can be processed by coda, as in

```
> summary(mcmc(X))
```

(where X is a (T, n) matrix whose n columns are the outputs of the MCMC algorithm for each component of the simulated object). Its derivative mcmc.list
is used to represent parallel runs of the same chain toward specific convergence
diagnoses. Specific multidimensional plots are also provided in levelplot,
acfplot, qqmath, densityplot, and xyplot. More specific diagnoses contained in coda will be discussed along with their generic description.

8.3 Monitoring convergence to stationarity

8.3.1 Graphical diagnoses

A first empirical approach to convergence control is to draw pictures of the
output of simulated chains, componentwise as well as jointly, in order to detect
deviant or nonstationary behaviors, with the obvious drawback that chains
"stuck" in a particular region of the parameter space, far away from the main
bulk of the target f, may well exhibit a stationary pattern. For instance, coda
provides this crude analysis via the plot command, which, when applied to
an mcmc object, produces both a trace of the chain across iterations and a
non-parametric estimate of its density, parameter by parameter (rather than
a standard plot).

Example 8.1. We again consider the random effect logit model already processed in Example 5.17 using a Monte Carlo EM strategy. Opting for a standard
random walk Metropolis–Hastings algorithm, we simulate both the random effects
u_i $(i = 1, \ldots, n)$ and the logit coefficient β from normal distributions centered
at the previous values of those parameters. (Note that the u_i's are independent
given the parameters β and σ.) The scale parameter σ of the random effect can
be simulated directly from an inverse gamma distribution. The core of the code
is then

```
> T=10^3
> beta=mlan
> sigma=1
> u=rnorm(n)*sigma
> samplu=matrix(u,nrow=n)
> for (iter in 2:T){
+   u=rnorm(n)
+   for (i in 1:n){
+     mu=samplu[i,iter-1]
+     u[i]=sigma[iter-1]*rnorm(1)+mu
```

```
+    if (log(runif(1))>gu(u[i],i,beta[iter-1],sigma[iter-1])
+         -gu(mu,i,beta[iter-1],sigma[iter-1]))
+       u[i]=mu}
+    samplu=cbind(samplu,u)
+    sigma=c(sigma,1/sqrt(2*rgamma(1,0.5*n)/sum(u^2)))
+    tau=sigma[iter-1]/sqrt(sum(as.vector(x^2)*
+         pro(beta[iter-1],u)))
+    betaprop=beta[iter-1]+rnorm(1)*tau
+    if (log(runif(1))>likecomp(betaprop,sigma[iter],u)
+      -likecomp(beta[iter-1],sigma[iter],u))
+       betaprop=beta[iter-1]
+ beta=c(beta,betaprop)
+ }
```

using some of the functions already defined in Example 5.17. Calling plot on
mcmc(cbind(beta,sigma)) then leads to the graphs presented in Figure 8.1. It
is clear from both traces (on the left) that both components of the chain move
very slowly over the range of possible values. As could have been predicted, both
components are, in addition, heavily correlated. This example is thus characteristic
of a case where plot is sufficient to identify convergence problems and signals
that 10^3 iterations are insufficient to ensure proper mixing.   ◀

Another natural graphical diagnosis that is appropriate both for this setting and for the convergence of averages is to check for the stabilization of
the empirical cdf derived from the Markov chains using for instance the function cumuplot of coda. Figure 8.2 illustrates the application of this function
for the MCMC sample of Example 8.1, with a clear lack of stability calling
for more simulations (or another sampler!). Note the sudden dip in the lower
2.5% quantile evaluation for the β sequence. It could correspond either to the
sudden discovery of another region supported by f or an outlying excursion
of the Markov chain. Given that the median sequence hardly moves at this
stage, the second explanation is more likely.

The graphs produced by coda are not easily open to calibration, as shown
by the poor rendering in Figure 8.2. This is especially true when monitoring
several parameters at the same time, as shown by plot(mcmc(t(samplu)))
in the example above. Note that using plot on an mcmc.list object (made
of the mcmc output of several chains run in parallel) produces a comparison
of the chains. Also, be warned that using cumuplot on a long MCMC chain
without first thinning out the chain is quite time-consuming since quantiles
are computed on the partial chain for each iteration!

8.3.2 Nonparametric tests of stationarity

If we want more confidence (than a mere graphical check) about the stationarity of the Markov chain $(x^{(t)})$, we need to check, in a statistical way, that

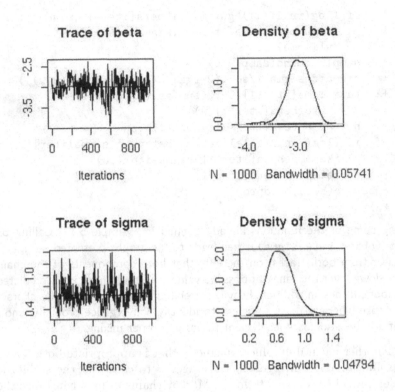

Fig. 8.1. Outcome of the `plot.mcmc` function applied to a sample of 10^3 values produced by a random walk Metropolis–Hastings algorithm based on a simulated dataset from the random effect logit model of Example 5.17 with $n = 20$, $m = 30$, $\beta = -3$, $\sigma = 1$, and the x_{ij}'s distributed uniformly over $\{-1, 0, 1\}$.

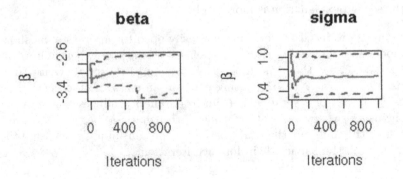

Fig. 8.2. Outcome of the `cumuplot` function applied to the same MCMC sample as in Figure 8.1. The lower plot corresponds to the lower 2.5% quantile, the central plot to the median, and the upper plot to the .975% quantile.

the distribution of the chain does not change over time. This can be done using a single chain or multiple chains, the setting being obviously easier on multiple chains given their independence.

For instance, standard non-parametric tests of fit, such as Kolmogorov–Smirnov or Cramer–von Mises tests, can be applied to a single output of the chain $(x^{(t)})$ to compare the distributions of the two halves (or other sub-parts) of this sample, $(x^{(1)}, \ldots, x^{(T/2)})$ and $(x^{(T/2+1)}, \ldots, x^{(T)})$. Since non-parametric tests are calibrated in terms of iid samples, there needs to be a correction for the Markov correlation between the $x^{(t)}$'s.

The solution is fairly simple and is also used to achieve the independence reproduction mentioned in Section 8.2.3. The correction can be achieved by introducing a *batch* size G and subsampling, leading to the construction of two (quasi-) independent samples. (Selecting G can be done along the way, for instance deducing G from the autocorrelation of the chain via acf or waiting until the stationarity test is accepted.) The corresponding subsamples $(x_1^{(G)}, x_1^{(2G)}, \ldots)$ and $(x_2^{(G)}, x_2^{(2G)}, \ldots)$ can be processed by ks.test in order to assess whether or not the *Kolmogorov–Smirnov statistic*

$$(8.2) \qquad K = \frac{1}{M} \sup_{\eta} \left| \sum_{g=1}^{M} \mathbb{I}_{(0,\eta)}(x_1^{(gG)}) - \sum_{g=1}^{M} \mathbb{I}_{(0,\eta)}(x_2^{(gG)}) \right|$$

is small enough. For multidimensional chains, (8.2) can be computed on either a function of interest or each component of the vector $x^{(t)}$.

Besides the static perspective of deciding whether or not a given number T of iterations is large enough, the statistic K can be processed to derive a stopping rule, computing the corresponding p-value using ks.test for a sequence of T's until it gets above a given level. A graphical indicator can be used as well, representing the sample of $\sqrt{M} \, K_T$'s against T and checking for stabilization.

Example 8.2. Considering again the nuclear pump failures of Example 7.12, we monitor the subchain $(\beta^{(t)})$ produced by the algorithm. Figure 8.3 *(top)* gives the value of the Kolmogorov–Smirnov p-values produced by ks.test as T increases to 10^3. This is a sequential graph in that each dot is based on the sample of $\beta^{(t)}$'s produced at time T and then thinned by a lag of $G = 10$ and separated into both halves. Note that the Gibbs sampler is initialized at the MLE of the λ_i's. Since Figure 8.3 *(top)* exhibits a lack of pattern and a fair spread of the p-value over $[0, 1]$, there is no indication of a lack of stationarity to deduce from this graph. If instead we compare two different chains with the same tool, Figure 8.3 *(bottom)* indicates a longer time pattern. Indeed, while the Kolmogorov–Smirnov p-values remain far away from low probabilities, the pattern exhibited in this comparison means that, while around iterations 4000 to 6000 both samples provide very similar empirical cdfs, they explore differently the space in the later iterations. Thus 10^4 iterations are not sufficient to provide a stable evaluation of the stationary distribution (for at least one chain). ◀

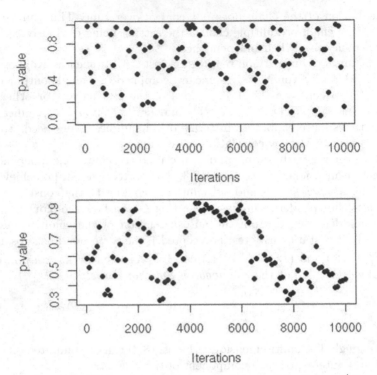

Fig. 8.3. Plot of 100 Kolmogorov–Smirnov p-values resulting from 10^4 iterations of the Gibbs sampler of Example 7.12 applied to the nuclear pump failure dataset. *(top)* Each dot corresponds to a Kolmogorov–Smirnov test applied to the two halves of the thinned sample of size T; *(bottom)* similar graph for two thinned parallel MCMC chains of size T.

We do not want to get into a debate about the p-value here, but note that, while its most objective definition is that it is uniformly distributed under the null hypothesis, that is rarely the case. Larger numbers of simulations from the Gibbs sampler still produce patterns as in Figure 8.3 *(bottom)*, even though both chains are clearly from the same distribution and close to independent for M large enough. For G extremely large, the p-values end up being almost always close to 1, which, following Jeffreys (1939), would mean rejecting the null hypothesis because of the lack of uniformity!

Exercise 8.2 Reproduce Figure 8.3 for two independent Gibbs samples of size 10^5, and evaluate the impact of the lag G on the shape of those graphs. (Try $G = 2, 5, 10, 100, 1000$.)

Heidelberger and Welch (1983) instead use the Cramer–von Mises statistic that approximates the L_2 distance between two distributions,

$$C = \int (F(x) - F_n(x))^2 \, \mathrm{d}F(x),$$

where F and F_n denote the reference cdf and the empirical cdf, respectively. In the case of two samples, $(x^{(1g)})_{1 \le g \le M}$ and $(x^{(2g)})_{1 \le g \le M}$, the corresponding comparison of the empirical cdfs is based on a normalized version of the rank variance. The test is implemented in the `coda` function `heidel.diag` as its first part (the function unfortunately does two tests at once). When used on the output of Example 8.2 (with 10^5 iterations), its output is

```
> heidel.diag(mcmc(beta))

        Stationarity start       p-value
        test           iteration
var1 passed            1         0.212
```

which complements the stationarity check brought by the Kolmogorov–Smirnov test.

↯ Figure 8.3 provides a forceful illustration of the fact that an assessment of stationarity based on a single chain is open to misinterpretation: In cases of strong attraction from a local mode, the chain will most likely behave as if it was simulated from the restriction of f to the neighborhood of this mode and thus lead to a positive convergence diagnosis (this is the *"you've only seen where you've been"* defect mentioned before). Using several chains with dispersed starting values is a precaution against this faulty conclusion, although not always foolproof if the dispersion is not adequate.

8.3.3 Spectral analysis

While we neither can nor want to get into theoretical details, there exist several practical convergence assessment devices based on *spectral analysis* (also known as Fourier analysis). For instance, Geweke (1992) uses the *spectral density* of $h(x^{(t)})$,

$$S_h(w) = \frac{1}{2\pi} \sum_{t=-\infty}^{t=\infty} \mathrm{cov}\left(h(x^{(0)}), h(x^{(t)})\right) e^{\iota t w},$$

where ι denotes the complex square root of 1 (that is, $e^{\iota t w} = \cos(tw) + \iota \sin(tw)$) to derive the asymptotic variance of (8.1) as $\gamma_h^2 = S_h^2(0)$. Estimating S_h by appropriate non-parametric methods, the `coda` function `geweke.diag` then constructs the equivalent of a t test to assess the equality of the means of the first and last parts of the Markov chain. More specifically, Geweke (1992) takes the first T_A and last T_B observations from a sequence of length T to derive

$$\delta_A = \frac{1}{T_A} \sum_{t=1}^{T_A} h(x^{(t)}), \qquad \delta_B = \frac{1}{T_B} \sum_{t=T-T_B+1}^{T} h(x^{(t)}),$$

and the estimates σ_A^2 and σ_B^2 of $S_h(0)$ based on both subsamples, respectively. The test statistic produced by `geweke.diag` is then the asymptotically normal so-called Z-score

$$\sqrt{T}(\delta_A - \delta_B) \Big/ \sqrt{\frac{\sigma_A^2}{\tau_A} + \frac{\sigma_B^2}{\tau_B}},$$

with $T_A = \tau_A T$, $T_B = \tau_B T$, and $\tau_A + \tau_B < 1$.

For instance, when used (again) on the output of Example 8.2 (with 10^5 iterations), Geweke's (1992) statistic is

```
> geweke.diag(mcmc(beta))

   var1
   0.2139
```

and it is thus compatible with stationarity. In the event `geweke.diag` returns a negative statement (that is, a value incompatible with a $\mathcal{N}(0, 1)$ distribution), `geweke.plot` can be used to see how much of the sample must be removed before achieving stationarity.

The function `geweke.diag` may sometimes result in an error message like

```
> geweke.diag(mcmc(xmc))
Error in data[1:nobs, , drop = FALSE] : incorrect number
of dimensions
```

without further explanation. It actually seems that the sample size is limited to $10^5 - 1$.

Another approach also using the spectral approximation to the asymptotic variance is found in Heidelberger and Welch (1983) and in the second half of `heidel.diag`, as in

```
> heidel.diag(mcmc(sigma))

       Halfwidth Mean Halfwidth
       test
var1 passed     1.05 0.00793
```

since the so-called *halfwidth* test uses the estimated asymptotic variance to normalize the difference between overall and partial means.

An ideal stationarity diagnosis would be to use an estimate of the "missing mass"; that is, of the area of the support of f that has not yet been explored by the MCMC chain. Unfortunately, that is very rarely the case! For one thing, the normalizing constant of the target f is almost never known in realistic situations. For another, if $\tilde{f} \propto f$, producing an estimate of

$$\int f(x)\,\mathrm{d}x$$

based on a sample simulated from f is much harder than it seems, and checking for the stabilization of this estimate faces the same difficulty as with other estimates, thanks to the *"you've only seen where you've been"* defect.

Exercise 8.3 Show that if φ is an arbitrary density with the same support as f, if $(x^{(t)})$ is a Markov chain with stationary distribution f, and if $f(x) = C\tilde{f}(x)$, then

$$\frac{1}{T}\sum_{t=1}^{T}\varphi(x^{(t)})/\tilde{f}(x^{(t)})$$

is converging (in T) to $1/C$. Discuss the constraints to impose upon φ to ensure this estimator has a finite variance.

As a final warning about the relative worth of convergence assessments, we provide below an example where every indicator gives a green light, except that the main mode of the target is missed by the Markov chain.

Example 8.3. Consider an AR(1) model $x_{t+1} = \varrho x_t + \epsilon_t$ $(t \geq 1)$, where $\epsilon_t \sim \mathcal{N}(0,1)$, observed through $y_t = x_t^2 + \zeta_t$, where $\zeta_t \sim \mathcal{N}(0,\tau^2)$. If we focus on the conditional distribution of x_t given (x_{t-1}, y_t, x_{t+1}), $\pi(x_t | x_{t-1}, y_t, x_{t+1})$, proportional to

$$\exp - \left\{ \tau^{-2}(y_t - x_t^2)^2 + (x_t - \rho x_{t-1})^2 + (x_{t+1} - \varrho x_t)^2 \right\}/2,$$

it is not a standard distribution and, to approximate it, we can run a random walk Metropolis–Hastings chain started at $\sqrt{y_t}$ (assuming $y_t > 0$). If the random walk relies on a scale of 0.1, both Geweke's (1992) and Heidelberger and Welch's (1983) criteria give a positive assessment of the MCMC sample xmc:

```
> geweke.diag(mcmc(xmc))

Fraction in 1st window = 0.1
Fraction in 2nd window = 0.5
    var1
-0.6162
> heidel.diag(mcmc(xmc))

        Stationarity start       p-value
```

```
        test            iteration
var1 passed            1              0.83

        Halfwidth Mean Halfwidth
        test
var1 passed       1.76 0.000697
```

Similarly, graphical tests all indicate a perfect fit to the target, as shown in the upper half of Figure 8.4. This is a special case where superimposing the true target on top of the histogram is possible because the dimension is one, π is known up to a normalizing constant, and this constant can be approximated by fitting the maximum of π with the maximum of a density approximation, as in

```
> ordin=apply(as.matrix(seq(min(xmc),max(xmc),le=200)),1,
+ FUN=function(x){exp(-.5*((xm*rho-x)^2+(x*rho-xp)^2+
+                 (yc-x^2)^2/tau^2))})
> lines(seq(min(xmc),max(xmc),le=200),
+ ordin*max(density(xmc)$y)/max(ordin))
```

The upper left-hand graph thus shows a perfect fit between the histogram of xmc and the target, while a Kolmogorov–Smirnov test applied to xmc thinned by a factor of $G = 50$ exhibits an almost periodic trend but no low p-value. If we now switch to another scale in the random walk, .9, both diagnoses conclude with a lack of stationarity and the Kolmogorov–Smirnov test represented on the lower right-hand side of Figure 8.4 gives a similar warning. However, when looking at the correspondance of the histogram and the target π, it is clear that the Markov chain has managed to escape the attraction of the lower mode to move to the most important mode. The lack of stationarity pointed out by the diagnostics is due to the initial values, but it also signals that proper coverage of both modes of the target π would require many more simulations than 10^5.   ◀

8.4 Monitoring convergence of averages

8.4.1 Graphical diagnoses

The initial and most natural diagnostic tool is to plot the evolution of the estimator (8.1) as T increases. If the curve of the cumulated averages has not stabilized after T iterations, the length of the Markov chain must be increased. The principle can be applied to multiple chains as well. The functions cumsum, plot.mcmc(coda), and cumuplot(coda) can be used for this purpose.

Example 8.4. (Continuation of Example 8.3) If we use several starting points located near either $\sqrt{y}_t$ or $-\sqrt{y}_t$, with a scale of 0.5,

```
> plot(mcmc.list(mcmc(smpl[,1]),...,mcmc(smpl[,M])))
```

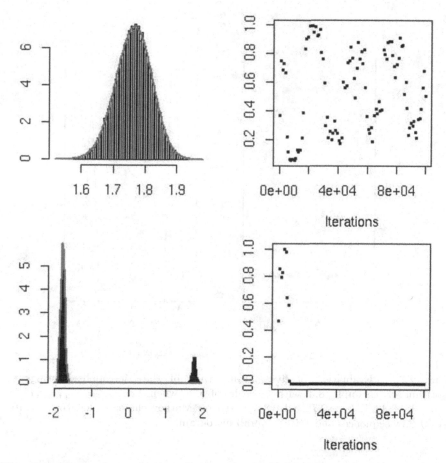

Fig. 8.4. Comparison of two scales in a random walk Metropolis–Hastings evaluation of the posterior distribution of x_t in the noisy AR model of Example 8.3 when $(x_{t-1}, y_t, x_{t+1}, \tau, \rho) = (-0.94, 3.17, -1.12, 0.2, 0.85)$ based on Markov chains of length $T = 10^5$. *(top)* Scale equal to 0.1 with histogram and fit *(left)* and Kolmogorov–Smirnov p-values; *(bottom)* same graphs for a scale of 0.9.

results in Figure 8.5, which clearly exhibits the problem with the convergence of the Metropolis–Hastings algorithm in this case. (The matrix smpl is made of the M parallel runs of the MCMC algorithm.) Note that using cumsum on a single chain started near $-x_t$ (and staying there) does not exhibit any deficiency because the chain remains stable in the vicinity of this secondary mode. ◀

Note the drawback, when using mcmc.list, in having to enter each separate simulation as mcmc(smpl[,i]). Defining smpl directly by smpl=c(smpl,mcmc(xmc)) in the loop for the parallel chains does not work.

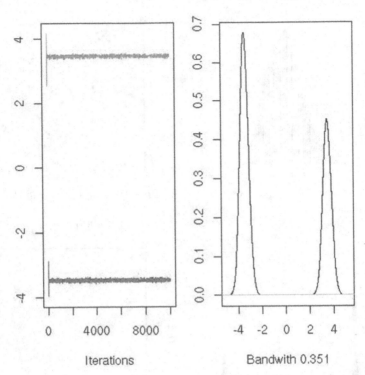

Fig. 8.5. Evaluation of five parallel runs of the Metropolis–Hastings algorithm of Example 8.3 with a scale of 0.5 when $(x_{t-1}, x_t, y_t, x_{t+1}, \tau, \rho) = (-3.13, -3.46, 12.02, -2.75, 0.2, 0.85)$ based on Markov chains of length $T = 10^4$. *(left)* Raw sequences and *(right)* overall histogram.

In addition, the density estimation in `cumuplot` is also based on the regular R `density` function and does not take into account the correlation between the observations, which means that the estimated density is usually spikier than it should be. (The same implication applies for `summary.mcmc`.) Most functions in `coda` are written in R, so it is also possible to check their contents and, if confident enough, modify them to correct, for instance, for the density estimation.

Thus, in most cases, the graph of either a single raw sequence $(x^{(t)})$ or the corresponding cumulated average is unhelpful in the detection of stationarity or convergence. Paradoxically, it is only when the chain has explored different regions of the state-space during the observation time that a lack of stationarity can be detected. (As in other testing setups, it is always easier to reject the stationarity assumption than to accept it.)

8.4.2 Within and between variances

Since multiple chains are almost always necessary to assess the convergence of the average, we now consider the central convergence tool of Gelman and Rubin (1992), implemented in `coda` as `gelman.diag(coda)` and `gelman.plot(coda)`. Using M parallel chains $\{x_m^{(t)}\}_t$ ($1 \leq m \leq M$, $1 \leq t \leq T$), possibly transformed into $\xi_m^{(t)} = h(x_m^{(t)})$, the stopping rule of Gelman and Rubin (1992) is based on the difference between a weighted estimator of the variance and the variance of estimators from the different chains.

The between-chain variance is defined as the variance of the averages

$$B_T = \frac{1}{M-1} \sum_{m=1}^{M} (\bar{\xi}_m - \bar{\xi})^2 \,,$$

while the within-chain variance is the average of the variances

$$W_T = \frac{1}{M-1} \sum_{m=1}^{M} s_m^2 = \frac{1}{M-1} \sum_{m=1}^{M} \frac{1}{T-1} \sum_{t=1}^{T} (\xi_m^{(t)} - \bar{\xi}_m)^2$$

with

$$\bar{\xi}_m = \frac{1}{T} \sum_{t=1}^{T} \xi_m^{(t)}, \qquad \bar{\xi} = \frac{1}{M} \sum_{m=1}^{M} \bar{\xi}_m \,.$$

Those quantities are commonly used in analyses of variance. A first estimator of the posterior variance of ξ is

$$\hat{\sigma}_T^2 = \frac{T-1}{T} W_T + B_T \,.$$

Since $\hat{\sigma}_T^2$ and W_T are asymptotically equivalent, Gelman and Rubin (1992) build their criterion on the comparison of those two quantities. As long as the different sequences $(\xi_m^{(t)})$ remain concentrated around their starting values (or on different portions of the parameter space), $\hat{\sigma}_T^2$ overestimates the variance of $\xi_m^{(t)}$ because of the large dispersion of the initial distribution, whereas W_T underestimates this variance. For instance, using the same five sequences as in Figure 8.5, the plot comparing both estimates, as provided by `gelman.plot`, is given in Figure 8.6. The criterion ("shrink factor") being very far from 1 shows, to no surprise, that the five chains have not yet covered (or not yet converged to) the same region. Note that the quantile in the graph is based on a t distribution, which amounts to assuming the target distribution to be close to normal. If this is clearly not the case, the chain should be transformed first (or the `transform=TRUE` option should be activated in `gelman.plot`, changing the chain via a logit transform).

The shrink factor of Gelman and Rubin (1992) is R_T, with

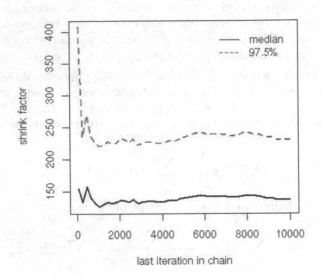

Fig. 8.6. Gelman and Rubin's (1992) evaluation of convergence for the noisy AR model and for the same chains as in Figure 8.5.

$$R_T^2 = \frac{\hat{\sigma}_T^2 + \dfrac{B_T}{M}}{W_T} \frac{\nu_T + 1}{\nu_T + 3},$$

where $\nu_T = 2(\hat{\sigma}_T^2 + \frac{B_T}{M})^2/W_T$ is the estimated degrees of freedom. While the approximate distribution of R_T^2 can be derived from an F approximation, $TB_T/W_T \sim \mathcal{F}(M - 1, 2W_T^2/\varpi_T)$ (see Gelman and Rubin, 1992, for details), with

$$\varpi_T = \frac{1}{M^2} \left[\sum_{m=1}^{M} s_m^4 - \frac{1}{M} \left(\sum_{m=1}^{M} s_m^2 \right)^2 \right],$$

`gelman.diag` and `gelman.plot` only report the value of the shrink factor, along with the 97.5% credible upper bound.

Example 8.5. (Continuation of Example 8.2) Figure 8.7 describes the evolution of the shrink factor based on ten parallel chains of $\beta^{(t)}$'s. It shows a clear stabilization around the target value 1 as early as 5000 iterations. The conclusion is thus in agreement with `geweke.diag` and `heidel.diag`.   ◀

This method has enjoyed wide use, in particular because of its simplicity and its intuitive connections with the standard tools of linear regression.

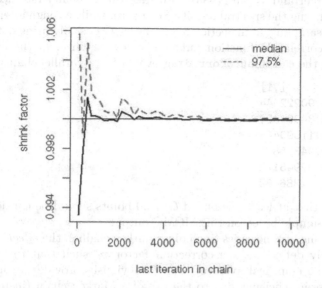

Fig. 8.7. Gelman and Rubin's (1992) evaluation for the pump failure model and for ten chains simulated as in Example 8.2.

However, we must warn you about the overconfidence it may induce. While the indicator R_T does converge to 1 under stationarity, its distributional approximation relies on normal assumptions, whereas the MCMC algorithms are used in settings where these approximations are at best difficult to satisfy and at worst not valid. The upper bound plotted by `gelman.plot` is therefore not necessarily an asymptotic .975 confidence band.

8.4.3 Effective sample size

In connection with Section 4.4, Section 8.3.3, and Section 8.2.3, we now consider the use of the effective sample size in the specific setting of MCMC algorithms, available as the `coda` function `effectiveSize`, because even within a stationary regime, there is an obvious difference between the number T of iterations and the size of an iid sample from f that would lead to the same variability. Indeed, the empirical average (8.1) cannot be associated with the standard variance estimator

$$\hat{\nu}_T = \frac{1}{T-1} \sum_{t=1}^{T} \left(h(x^{(t)}) - S_T \right)^2$$

due to the correlations amongst the $x^{(t)}$'s. The solutions of Geweke (1992) and Heidelberger and Welch (1983) both take this difficulty into account by directly estimating the spectral density S_h at zero, while a rough alternative is to rely on subsampling, as in Section 8.3.2, the lag G possibly being determined via the autocorrelation function `autocorr`. For instance, in the setting of Example 8.5, the call to `autocorr.diag` over the ten parallel chains returns

```
            [,1]
Lag 1  0.90222644
Lag 10 0.34901530
Lag 20 0.11590444
Lag 30 0.04325814
Lag 40 0.02945172
Lag 50 0.02386088
```

which means that at least one out of $G = 30$ points should be considered for a pseudo-iid sample based on this MCMC output.

In the setting of time series (including Markov chains), the *effective sample size* is directly defined as the correction factor τ_T such that $\hat{\nu}_T/\tau_T$ is the variance of the empirical average (8.1). It obviously provides an indication about the loss in efficiency due to the use of a Markov chain (instead of an iid sample) and can be computed as in Geweke (1992) and Heidelberger and Welch (1983) by

$$\tau_T = T/\kappa(h),$$

where $\kappa(h)$ is the autocorrelation time associated with the sequence $h(x^{(t)})$,

$$\kappa(h) = 1 + 2 \sum_{t=1}^{\infty} \text{corr}\left(h(x^{(0)}), h(x^{(t)})\right),$$

estimated by `spectrum0` under `coda`. Using `effectiveSize` over the ten parallel chains treated above leads to an effective sample size of 2645.168, meaning a 5% efficiency, which is somehow coherent with the $G = 30$ lag found above. When applied to a single sequence of 10^3 values $(\beta^{(t)}, \sigma^{(t)})$ from Example 8.1, the outcome is

```
> effectiveSize(mcmc(cbind(beta,sigma)))
    beta    sigma
 55.3948 236.0124
```

which again agrees with the heuristic message of Figure 8.1.

If you explore `coda` further, you will notice that it includes another diagnostic function called `raftery.diag`. While this function produces very explicit evaluations of the number of simulations to use to estimate a given quantile of the target distribution with a given precision, the theoretical foundations of this method due to Raftery and Lewis (1992) are not sound: The underlying structure is not a Markov chain (see Robert and Casella, 2004, Section 12.4.1).

Exercise 8.4 Produce several parallel MCMC chains $\{\beta_m^{(t)}\}_{t,m}$ for the pump data model as in Example 8.2 and study the variability of the number of simulations N proposed by `raftery.diag`. Compare this with the other stopping rules contained in coda.

8.4.4 Fixed-width batch means

The previous section indicated a rather rudimentary way to implement the requirement of Section 8.2.3, which is to replace the original Markov chain produced by an MCMC algorithm with a pseudo-iid sequence, namely by subsampling according to the rate given by the effective sample size. It is, however, feasible to study the convergence of an MCMC sequence by looking at the approximation error of the MCMC average

$$\delta_T = \frac{1}{T} \sum_{t=1}^{T} h(x^{(t)}).$$

Section 8.3.3 produced an estimator of the asymptotic variance of δ_T based on spectral analysis, but there exist alternatives based on batch sampling and the Central Limit Theorem, such as the one introduced by Jones et al. (2006), which compares with `gelman.diag` (Flegal et al., 2008).

Batch sampling was used in Section 8.3.2 for deciding about the stationarity of the MCMC sequence, but it is also possible to use this technique for convergence assessment. If batch means are defined by

$$\bar{h}_j = \frac{1}{a_T} \sum_{t=j}^{j+a_T-1} h(x^{(t)}),$$

and if the sequence a_T is increasing slowly enough to infinity, $a_T = \lfloor T^\nu \rfloor$ with $1/2 \le \nu \le 3/4$, then the approximation of the asymptotic variance of δ_T,

$$\hat{\sigma}_h^2 = \frac{T a_T}{(T - a_T)(T - a_T - 1)} \sum_{t=1}^{T-a_T+1} (\bar{h}_t - \delta_T)^2,$$

converges to the asymptotic variance with T (under some extra regularity conditions on the chain $(x^{(t)})$). While only validated by asymptotics (i.e., when T goes to infinity), and therefore open to misbehaving[2] for a fixed T, this result of Jones et al. (2006) allows a quick assessment of the variation of δ_T and even for a stopping rule based on the range $t_{1-\alpha/2}\hat{\sigma}_h/\sqrt{T}$ being smaller than a fixed precision for a confidence range of $(1 - \alpha)$. Jones et al. (2006) also mention the possibility of bootstrapping the $\bar{h}_j$'s to construct the confidence interval on $\mathbb{E}_f[H]$.

---

[2] When a_T is fixed, $\hat{\sigma}_h^2$ is not a convergent estimator of the asymptotic variance σ_h^2, as shown by Glynn and Whitt (1992).

Example 8.6. (Continuation of Example 8.5) If we apply this approach to the chain $(\beta^{(t)})$ produced for the pump failure data, computing fixed-width batch means using the sequence $a_t = \lfloor T^{5/8} \rfloor$, we obtain the convergence diagnostic illustrated by Figure 8.8. The computation of the 25 confidence ranges on which this picture is built is the result of the R code

```
> Ts=seq(100,T,le=25)
> ranj=matrix(0,ncol=2,nrow=25)
> for (j in 1:25){
+ aT=trunc(Ts[j]^(5/8))
+ batch=rep(mean(beta),le=Ts[j]-aT)
+ for (t in (aT+1):Ts[j]) batch[t-aT]=mean(beta[(t-aT):t])
+ sigma=2*sqrt(sum((batch-mean(beta[1:Ts[j]]))^2)*aT/
+   ((Ts[j]-aT)*(Ts[j]-aT-1)))
+ ranj[j,]=mean(beta[1:Ts[j]])+c(-sigma,+sigma)
+ }
```

Note that the confidence band decreases much faster than a standard confidence band for an iid series.   ◀

Exercise 8.5 Build an alternative confidence band on the MCMC output of Example 8.6 by deriving the variance estimates via a boostrap evaluation based on the batch means, and compare it with the band of Figure 8.8.

8.5 Adaptive MCMC

8.5.1 Cautions about adaptation

Now that convergence diagnoses have been made available to you, we can consider the next step in constructing MCMC algorithms, which is to calibrate (and thus rank) them according to their performance against those diagnoses. This is the area of *adaptive* algorithms, a recently developed extension of MCMC algorithms where the transition kernels are tuned on-line according to the performances observed so far. The goal of using the "best" kernel within a collection of proposals or the "best" parameters within a family of parameterized kernels—such as, for example, random walks with unidimensional or multidimensional scales—is, however, difficult to implement in practice, if only because of the potentially lethal impact on the convergence properties of the corresponding algorithm!

Indeed, if we keep tuning the algorithm according to its outcome until the present time, it means that the algorithm is no longer Markovian since it depends on the entire past of the simulation history. There are therefore severe constraints to be put on the adaptation process if you want to preserve the

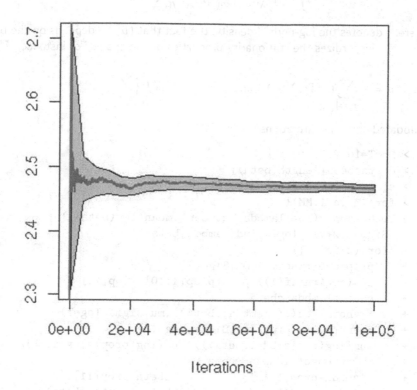

Fig. 8.8. Evolution of a cumulated MCMC average for the pump failure model of Example 8.2. The grey background corresponds to the 95% confidence range estimated by the fixed-width batch sampling estimate of Jones et al. (2006).

(theoretical) insurance that your algorithm will converge. While the theory behind adaptive MCMC algorithms is beyond our reach in this book, we primarily want to point out the danger of naïve MCMC adaptation before discussing the existing amcmc package developed by Rosenthal (2007).

Example 8.7. For the pump failure model of Example 8.2, instead of using the Gibbs steps, we could try to build an independent log-normal proposal with parameters (μ, Σ) based on the empirical moments of the Markov chain. The intuition behind this construction would be to approximate the true target by a (log-)normal distribution with the same moments as the target distribution, assuming the empirical moments constructed at time t provide convergent approximations to the true moments. Unfortunately, even if we accept the proposed value $(\beta^\star, \lambda^\star)$ with the standard Metropolis–Hastings ratio

$$\frac{f(\beta^\star, \lambda^\star)}{f(\beta^{(t)}, \lambda^{(t)})} \; \frac{\varphi(\beta^{(t)}, \lambda^{(t)}|\mu, \Sigma)}{\varphi(\beta^\star, \lambda^\star|\mu, \Sigma)} \wedge 1,$$

where φ denotes the log-normal density, the fact that (μ, Σ) depends on the past iterations jeopardizes the stationarity properties of the chain. For instance, if

$$\mu = \frac{1}{t} \sum_{m=1}^{t} (\beta^{(m)}, \lambda^{(m)}) \quad \text{and} \quad \Sigma = 0.2 \, \text{var}\left(\left\{\beta^{(m)}, \lambda^{(m)}\right\}_{1 \leq m \leq t}\right)$$

is updated every 100 iterations, as in

```
> MM=T=10^2
> cbeta=beta[length(beta)]
> clambda=lambda[length(beta),]
> for (m in 1:MM){
+   mu=c(apply(log(lambda),2,mean),mean(log(beta)))
+   Sigma=.2*var(log(cbind(lambda,beta)))
+   for (t in 1:T){
+     prop=exp(rmnorm(1,mu,Sigma))
+     if (log(runif(1))>post(prop[1:10],prop[11])-
+     post(clambda,cbeta)+
+     dmvnorm(log(c(clambda,cbeta)),mu,Sigma,log=T)-
+     dmvnorm(log(prop),mu,Sigma,log=T)-
+     sum(log(c(clambda,cbeta)))+sum(log(prop))) #jacobian
+         prop=c(clambda,cbeta)
+     clambda=prop[1:10];           cbeta=prop[11]
+     lambda=rbind(lambda,clambda);beta=c(beta,cbeta)
+   }}
```

(which follows $T = 10^2$ regular iterations of the Gibbs sampler that produce beta and lambda, and thus the first estimates μ and Σ), then the chain $(\beta^{(t)}, \lambda^{(t)})$ collapses, as shown by Figure 8.9 for $\beta^{(t)}$. Obviously, there is no particular reason to pick the shrinkage factor in Σ that way, but our point in this example is to show that using a perfectly valid Metropolis–Hastings step does not always work when the parameters of the proposal keep being updated. ◀

Exercise 8.6 Repeat the experiment conducted in Example 8.7 when Σ is updated as

```
Sigma=var(log(cbind(lambda,beta)))
```

and contrast the density estimate of the posterior on β thus obtained with an estimate resulting from a Gibbs sample as in Example 8.2 and Figure 5.17.

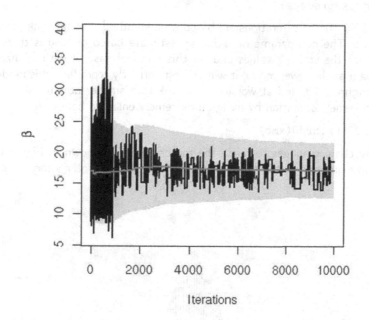

Iterations

Fig. 8.9. A degenerating MCMC adaptation based on an independent proposal and current moments for the pump failure model of Example 8.2. The grey background corresponds to the normal range offered by the adaptation, the raw curve is one realization of the adaptive MCMC, and the lighter curve on top is the sequence of averages of the normal proposals.

Even more elaborate learning scenarios may fall short from converging, as shown by the following example.

Example 8.8. (Continuation of Example 8.4) If we use an early MCMC sample to build a non-parametric estimate of the target density in the shape of a normal mixture centered at each point in the sample, with a common standard deviation provided by `bw.nrd(xmc)`, we can keep updating the non-parametric estimate with each new value of the corresponding Markov chain. The R code thus looks like

```
> for (t in (T+1):TT){
+ bw=bw.nrd0(xmc)
+ prop=rnorm(1,mean=sample(xmc,1),sd=bw)
+ prodens=log(density(xmc,from=prop,to=prop,n=1,bw=bw)$y)
+ if ((is.na(prop))||(log(runif(1))>ef(prop)-ef(xmc[t-1])+
+       curdens-prodens)){
+    prop=xmc[t-1]
```

```
+     prodens=curdens}
+ xmc=c(xmc,prop)
+ curdens=prodens}
```

where the $T = 10^2$ first iterations are based on a standard random walk proposal to build xmc. The non-parametric density estimate based on xmc is then very dependent on the starting values and, in this bimodal case, if we initialize the Markov chain at the lower mode, it will at best correctly reproduce this mode, as shown in Figure 8.10, and at worst even shrink to a smaller domain. If we now modify the kernel estimation by using an extremely enlarged bandwidth,

```
> bw=500*bw.nrd0(xmc)
```

the Markov chain manages to reach the major mode, as shown in Figure 8.11, but the approximation of the target distribution is still not satisfactory after 10^5 iterations.   ◀

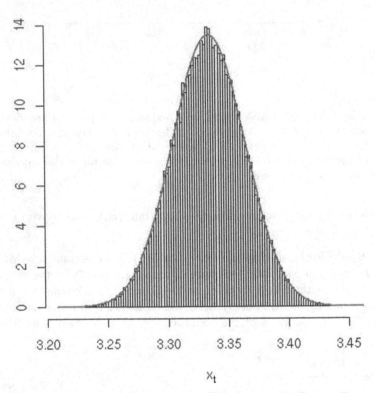

Fig. 8.10. Output of a non-parametric MCMC adaptation based on a kernel estimate of the target density and 10^4 iterations for the noisy AR model against the secondary mode of the target density.

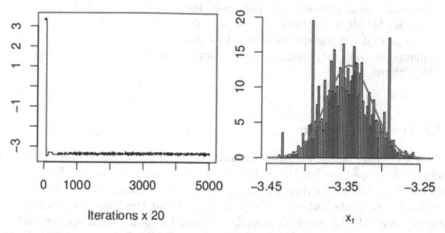

Fig. 8.11. Recovery of the main mode of the target density for the noisy AR model when using a non-parametric MCMC adaptation based on a kernel estimate with a very large bandwidth estimate. *(left)* Rawplot of the Markov chain; *(right)* fit of the histogram against the target.

The usual drawback of adaptive methods is that they rely *too much* on past samples and thus emphasize the blind angles of those samples, as Example 8.7 clearly expresses. While a regular Metropolis–Hastings algorithm may find sufficient energy to overcome this attraction of the local modes, an adaptive version will find itself mired much more often in this situation, if not always.

The solution found in the literature for this difficulty is to progressively tone/tune down the adaptive aspect. More precisely, Roberts and Rosenthal (2009) propose a *diminishing adaptation* condition that states that the total variation distance between two consecutive kernels must uniformly decrease to zero (which does not mean that the kernel must converge!). For instance, a random walk proposal that relies on the empirical variance of the sample (modulo a ridge-like stabilizing factor) as in Haario et al. (1999) will satisfy this condition. Another possibility found in Roberts and Rosenthal (2009) is to tune the scale in each direction toward an optimal acceptance rate of 0.44, which is the solution implemented in the amcmc package, described below. More precisely, for each component of the simulated parameter, a factor δ_i corresponding to the logarithm of the random walk standard deviation is updated every 50 iterations by adding or subtracting a factor ϵ_t depending on whether or not the average acceptance rate on that batch of 50 iterations and for this component was above or below 0.44. If ϵ_t decreases to zero as $\min(.01, 1/\sqrt{t})$, the conditions for convergence are satisfied. (This is exactly the rule followed by amcmc.)

⚡ We must stress that we only skim over the conditions required for an adaptive MCMC algorithm to converge, mainly to introduce the following package that has algorithms known to converge. You must refer to the appropriate (difficult) literature if you want to validate your own adaptive algorithms!

8.5.2 The amcmc package

Let us start with a few words of warning about the unusual aspects of this package, developed by Jeff Rosenthal of the University of Toronto. First, the package does not come through the usual CRAN depository but must be downloaded from the author's Webpage (at least at the time this book was written). Second, the program must be started by calling source("amcmc") if the corresponding file amcmc is available in the local directory. Third, it requires the C compiler gcc to be available, as amcmc executes part (or all) of the computation using a C code compiled by gcc. When the target density f and possibly the target functional h can be programmed in C as well, Rosenthal (2007) shows that the execution time can be improved one hundredfold compared with an R implementation.

⚡ Due to the C compiler requirements of the amcmc package, the R code corresponding to the following examples has not been included in our demo(Chapter.8) program. Note, however, that Example 8.9 is already provided within the amcmc package, while both following examples need to be reprogrammed.

These preliminaries having been stated, the program can be tested on any example, even though it may take a while to execute.

Example 8.9. The resident example provided with amcmc is a famous baseball dataset used in the James–Stein estimation literature (Efron and Morris, 1975). It is an analysis of variance model where observables Y_i ($1 \le i \le 18$) are normally distributed, $Y_k \sim \mathcal{N}(\theta_i, \sigma^2)$, the μ_i's being also iid normal, $\theta_i \sim \mathcal{N}(\mu, \alpha)$, with the common mean being distributed as $\mu \sim \mathcal{N}(0, 1)$ and the scale as $\alpha \sim \mathcal{IG}(2, 2)$. In this example, σ^2 is replaced with its empirical Bayes estimate, equal to 0.00434. Calling

```
> amcmc("baseballlogdens","baseballfirstfunct",logflag=TRUE)
[1] 0.3931286
```

thus produces 0.393 as the Bayes estimate of θ_1 in about one minute. Using instead the C version

```
> cfns("baseball")
gcc-4.2 -std=gnu99 -I/usr/share/R/include -I/usr/share/R/
include -fpic  -g -02 -c baseball.c -o baseball.o
gcc-4.2 -std=gnu99 -shared   -o amcmc.so amcmc.o baseball.o
-L/usr/lib/R/lib -1R
> amcmc("baseballlogdens","baseballfirstfunct",log=T,cfn=T)
[1] 0.3917384
```

leads to the same value in about one second!   ◀

> **Exercise 8.7** Show that the posterior distribution on μ in Example 8.9 can be obtained in a closed form, and take advantage of this availability to check the performances of a standard Gibbs sampler when using geweke, heidel, and the Kolmogorov–Smirnov assessment of Section 8.3.2.

The amcmc package can be used in conjunction with the coda convergence assessment tools when adding the option write=T, which saves the output in the amcmcvals file and the log proposal variances after each batch in the amcmcsigmas file. Both files are stored as R lists, with components values, a matrix whose rows correspond to the components of the vector to be simulated and whose columns correspond to the MCMC iterations. This means they can be uploaded as source("amcmcvals") and then checked via coda as in, for instance, summary(mcmc(t(amcmcvals))).

Example 8.10. (Continuation of Example 8.7) For the nuclear pump failure data model, the only modification required on the definition of the posterior distribution is a change of parameterization. Indeed, both β and the λ_i's being positive parameters, the random walk proposal used in amcmc requires the use of a log parameterization on the model, which means defining

```
apost=function(lambdabeta){
 lambda=lambdabeta[1:10]
 beta=lambdabeta[11]
 return((-exp(beta)*delta)+(((10*alpha)+gamma)*beta)+
 sum((-(time+exp(beta))*exp(lambda))+((data+alpha)*lambda)))
 }
```

as the substitute density (note the Jacobian term log(lambda)+beta in this function). When calling

```
> amcmc(dens=apost,func=mytarg,logfl=TRUE,num=10^4,vec=11)
[1] 2.491654
```

(where vec is the abbreviation of vectorlength, which provides the number of components of density), it quite compares with the expectation obtained from a regular Gibbs run,

```
> mean(beta)
[1] 2.487719
```

where beta is the result of $T = 5$ $times10^4$ Gibbs iterations. As noted above, the outcome of the simulation can be checked by coda, as for instance in

```
> source("amcmcvals")
> effectiveSize(mcmc(t(amcmcvals)))
    var1     var2     var3     var4     var5     var6     var7
  9411.0   9308.9  10001.0  10001.0   9357.6  10102.4   8867.6
    var8     var9    var10    var11
  9253.3   9174.5   9495.4   8106.1
```

(but be warned that the theoretical validation of those diagnoses is necessarily limited due to the adaptive nature of the underlying Markov chain).   ◀

⚡A drawback of using amcmc when compared with standard R packages is that it requires the C compiler gcc, and, as a result, does not run under Windows (which of course is only a drawback when you are using Windows!). The positive side and the reason why we do include amcmc in the book are that it provides a correct adaptive MCMC algorithm that applies to any target π and in addition gives an insight on the underexploited possibilities of linking R with C and then gaining several orders of magnitude in computing time.

Obviously, adaptivity as implemented in amcmc does not solve all convergence issues, as shown below in the case of the noisy AR model of Example 8.3.

Example 8.11. (Continuation of Example 8.8) When using amcmc on the posterior distribution ef

```
> ef=function(x){
+      -.5*((xm*rho-x)^2+(x*rho-xp)^2+(yc-x^2)^2/tau^2
+      }
> amcmc(dens="eef",function="firstcoord",init=sqrt(yc),log=T)
[1] 2.900830
```

the final estimate of x_t is 2.9, located on the secondary mode of the posterior, while the simulated value is $x_t = -2.896$ (with $y_t = 8.49$, $x_{t-1} = -3.795$, $x_{t+1} = -2.603$).   ◀

8.6 Additional exercises

Exercise 8.8 The witch's hat distribution

$$\pi(\theta|y) \propto \left\{(1-\delta)\, \sigma^{-d} e^{-\|y-\theta\|^2/(2\sigma^2)} + \delta\right\} \mathbb{I}_C(\theta), \qquad y \in \mathbb{R}^d,$$

when θ is restricted to the unit cube $C = [0,1]^d$, has been proposed by Matthews (1993) as a calibration *benchmark* for MCMC algorithms.

a. Construct an algorithm that correctly simulates the witch's hat distribution. (*Hint:* Show that direct simulation is possible.)
b. The choice of δ, σ, and d can lead to arbitrarily small probabilities of either escaping the attraction of the mode or reaching it. Find sets of parameters (δ, σ, y) for which these two phenomena occur.
c. In each case, test the various convergence diagnoses proposed by coda to check whether or not they notice the difficulty.

Exercise 8.9 Consider the generator associated with the Markov chain $(X^{(t)})$ such that

$$(8.3) \qquad X^{(t+1)} = \begin{cases} Y \sim \mathcal{B}e(\alpha+1,1) & \text{with probability } x^{(t)} \\ x^{(t)} & \text{otherwise} \end{cases}$$

a. Show that the Markov chain is associated with the stationary distribution

$$f(x) \propto x^{\alpha+1-1}/\{1-(1-x)\}\,\mathbb{I}_{(0,1)}(x) = x^{\alpha-1}\,\mathbb{I}_{(0,1)}(x),$$

that is, with the beta distribution $\mathcal{B}e(\alpha, 1)$.
b. Apply the entire range of convergence diagnoses proposed in this chapter to the first 10^4 values of the Markov chain when $\alpha = .2$. Compare this with the outcome based on the first 10^6 values.
c. Reproduce the analysis when looking at a Metropolis–Hastings algorithm based on the same beta proposal $\mathcal{B}e(\alpha+1,1)$. (*Hint:* Show that y is then accepted with probability $x^{(t)}/y$.)

Exercise 8.10 Reproduce the convergence analysis of Example 8.1 when the logit structure is replaced with a probit dependence in the generalized linear model. Take advantage of the normal latent variables inherent in the probit model to compare the convergence properties of the algorithm inspired by Example 8.1 with a Gibbs sampler based on the normal latent variables.

Exercise 8.11 Consider the posterior distribution associated with the standard probit model

$$\prod_{i=1}^{n} \Phi(r_i\beta/\sigma)^{d_i} \Phi(-r_i\beta/\sigma)^{1-d_i} \times \pi(\beta, \sigma^2),$$

where

$$\pi(\beta, \sigma^2) = \sigma^{-4}\, \exp\{-1/\sigma^2\}\, \exp\{-\beta^2/50\},$$

and the pairs (r_i, d_i) are the observations, taken from Pima.tr as Pima.tr\$ped for r_i and Pima.tr\$type for y_i.

a. Build an R program that simulates from this posterior distribution based on a simple Gibbs sampler where β and σ^2 are alternatively simulated by normal and log-normal random walk proposals and accepted by a one-dimensional Metropolis–Hastings step.
b. Evaluate the convergence of this algorithm using multiple chains and gelman.diag.
c. Compare the convergence of this algorithm with that of a traditional Gibbs sampler based on the completion of the model using the normal latent variables associated with the normal cdf.

Exercise 8.12 Consider the target density

$$f(x) = \frac{\exp{-x^2/2}}{\sqrt{2\pi}} \frac{4(x - .3)^2 + .01}{4(1 + (.3)^2) + .01}.$$

a. Show that f integrates to 1 and that it is a bimodal density.
b. Implement a normal random walk Metropolis–Hastings algorithm with a small variance like .04, and use plot.mcmc, cumuplot, and heidel.diag to assess the convergence when starting from $x = -2$ and $x = 2$.
c. Compare those assessments with an on-line evaluation of the integral $\int f(x)\,dx$ based on the MCMC sample thus produced.

Exercise 8.13 In the setting of Example 8.2, find a thinning lag G large enough that the distribution of the Kolmogorov–Smirnov p-values has no visible pattern.

Exercise 8.14 Evaluate the impact of parameterization on gelman.diag for the model of Example 8.2 using the same MCMC sample on each case.

Exercise 8.15 (Tanner, 1996) Show that if $\theta^{(t)} \sim \pi^t$ and if the stationary distribution is the posterior density associated with $f(x|\theta)$ and $\pi(\theta)$, the weight

$$\omega_t = \frac{f(x|\theta^{(t)})\pi(\theta^{(t)})}{\pi^t(\theta^{(t)})}$$

converges to the marginal $m(x)$.

Exercise 8.16 As a summary exercise:

a. Build an R function that takes as input a (T, p) matrix representing the output of an MCMC run with T iterations and p components and that produces as its output p spreadsheets (one for every component) made of raw plots, density estimates, autocorrelation functions, Kolmogorov-Smirnov tests, Geweke's plots, and outcomes of cumuplot on a single page.
b. Parameterize the function such that those spreadsheets are automatically saved in an open format (pdf, eps, jpg, etc.).

Exercise 8.17 Recycling the examples used in this chapter, study whether a graphical plot of the evolution of the effective sample size across iterations is a trustworthy tool for assessing convergence.

Exercise 8.18 Since the true target in Exercise 8.9 is known, study the convergence pattern for an adaptive Metropolis–Hastings step whose proposal is a beta $\mathcal{Be}(\alpha_t + 1, 1)$ distribution with parameter α_t updated as the average over the previous iterations.

References

Albert, J. (2009). *Bayesian Computation with R*, second edition. Springer–Verlag, New York.

Beaumont, M., Zhang, W., and Balding, D. (2002). Approximate Bayesian computation in population genetics. *Genetics*, 162:2025–2035.

Berger, J., Philippe, A., and Robert, C. (1998). Estimation of quadratic functions: reference priors for non-centrality parameters. *Statistica Sinica*, 8(2):359–375.

Besag, J. and Clifford, P. (1989). Generalized Monte Carlo significance tests. *Biometrika*, 76:633–642.

Booth, J. and Hobert, J. (1999). Maximizing generalized linear mixed model likelihoods with an automated Monte Carlo EM algorithm. *Journal of the Royal Statistical Society Series B*, 61:265–285.

Boyles, R. (1983). On the convergence of the EM algorithm. *Journal of the Royal Statistical Society Series B*, 45:47–50.

Brooks, S. and Roberts, G. (1998). Assessing convergence of Markov chain Monte Carlo algorithms. *Statistics and Computing*, 8:319–335.

Cappé, O., Douc, R., Guillin, A., Marin, J.-M., and Robert, C. (2008). Adaptive importance sampling in general mixture classes. *Statistics and Computing*, 18:447–459.

Casella, G. (1996). Statistical theory and Monte Carlo algorithms (with discussion). *TEST*, 5:249–344.

Casella, G. and Berger, R. (2001). *Statistical Inference*, second edition. Wadsworth, Belmont, CA.

Casella, G. and George, E. (1992). An introduction to Gibbs sampling. *The American Statistician*, 46:167–174.

Casella, G. and Robert, C. (1996). Rao-Blackwellisation of sampling schemes. *Biometrika*, 83(1):81–94.

Casella, G. and Robert, C. (1998). Post-processing accept–reject samples: recycling and rescaling. *J. Comput. Graph. Statist.*, 7(2):139–157.

Chen, M., Shao, Q., and Ibrahim, J. (2000). *Monte Carlo Methods in Bayesian Computation*. Springer–Verlag, New York.

C.P. Robert, G. Casella, *Introducing Monte Carlo Methods with R*, Use R,
DOI 10.1007/978-1-4419-1576-4, © Springer Science+Business Media, LLC 2010

Chib, S. (1995). Marginal likelihood from the Gibbs output. *Journal of the American Statistical Association*, 90:1313–1321.

Cowles, M. and Carlin, B. (1996). Markov chain Monte Carlo convergence diagnostics: a comparative study. *Journal of the American Statistical Association*, 91:883–904.

Crawley, M. (2007). *The R Book*. John Wiley, New York.

Dalgaard, P. (2002). *Introductory Statistics with R*. Springer–Verlag, New York.

Del Moral, P. and Miclo, L. (1999). On the convergence and applications of generalized simulated annealing. *SIAM Journal on Control and Optimization*, 37(4):1222–1250.

Dempster, A., Laird, N., and Rubin, D. (1977). Maximum likelihood from incomplete data via the EM algorithm (with discussion). *Journal of the Royal Statistical Society Series B*, 39:1–38.

Dickey, J. (1968). Three multidimensional integral identities with Bayesian applications. *Annals Mathematical Statistics*, 39:1615–1627.

Doucet, A., Godsill, S., and Robert, C. (2002). Marginal maximum a posteriori estimation using Markov chain Monte Carlo. *Statistics and Computing*, 12:77–84.

Efron, B. and Morris, C. (1975). Data analysis using Stein's estimator and its generalizations. *Journal of the American Statistical Association*, 70:311–319.

Efron, B. and Tibshirani, R. (1993). *An Introduction to the Bootstrap*. Chapman and Hall, New York.

Evans, M. and Swartz, T. (2000). *Approximating Integrals via Monte Carlo and Deterministic Methods*. Oxford University Press, Oxford.

Feller, W. (1971). *An Introduction to Probability Theory and its Applications*, volume 2. John Wiley, New York.

Flegal, J., Haran, M., and Jones, G. (2008). Markov chain Monte Carlo: can we trust the third significant figure? *Statistical Science*, 23(2):250–260.

Gaetan, C. and Yao, J.-F. (2003). A multiple-imputation Metropolis version of the EM algorithm. *Biometrika*, 90:643–654.

Gaver, D. and O'Muircheartaigh, I. (1987). Robust empirical Bayes analysis of event rates. *Technometrics*, 29:1–15.

Gelfand, A. and Dey, D. (1994). Bayesian model choice: asymptotics and exact calculations. *Journal of the Royal Statistical Society Series B*, 56:501–514.

Gelfand, A. and Smith, A. (1990). Sampling based approaches to calculating marginal densities. *Journal of the American Statistical Association*, 85:398–409.

Gelman, A. and Hill, J. (2006). *Data Analysis Using Regression and Multi-level/Hierarchical Models*. Cambridge University Press, Cambridge.

Gelman, A. and Rubin, D. (1992). Inference from iterative simulation using multiple sequences (with discussion). *Statistical Science*, 7:457–511.

Geman, S. and Geman, D. (1984). Stochastic relaxation, Gibbs distributions and the Bayesian restoration of images. *IEEE Transaction on Pattern Analysis and Machine Intelligence*, 6:721–741.

Gentle, J. E. (2002). *Elements of Computational Statistics.* Springer–Verlag, New York, New York.

Genz, A. and Azzalini, A. (2009). *mnormt: The multivariate normal and t distributions.* R package version 1.3-3.

Geweke, J. (1992). Evaluating the accuracy of sampling-based approaches to the calculation of posterior moments (with discussion). In Bernardo, J., Berger, J., Dawid, A., and Smith, A., editors, *Bayesian Statistics 4*, pages 169–193. Oxford University Press, Oxford.

Geyer, C. and Thompson, E. (1992). Constrained Monte Carlo maximum likelihood for dependent data (with discussion). *Journal of the Royal Statistical Society Series B*, 54:657–699.

Glynn, P. W. and Whitt, W. (1992). The asymptotic validity of sequential stopping rules for stochastic simulations. *Annls of Applied Probability*, 2:180–198.

Haario, H. and Sacksman, E. (1991). Simulated annealing in general state space. *Advances in Applied Probability*, 23:866–893.

Haario, H., Saksman, E., and Tamminen, J. (1999). Adaptive proposal distribution for random walk Metropolis algorithm. *Computational Statistics*, 14(3):375–395.

Hàjek, B. (1988). Cooling schedules for optimal annealing. *Mathematics of Operations Research*, 13:311–329.

Hastings, W. (1970). Monte Carlo sampling methods using Markov chains and their application. *Biometrika*, 57:97–109.

Heidelberger, P. and Welch, P. (1983). A spectral method for confidence interval generation and run length control in simulations. *Communications of the Association for Computing Machinery*, 24:233–245.

Hesterberg, T. (1995). Weighted average importance sampling and defensive mixture distributions. *Technometrics*, 37:185–194.

Jacquier, E., Johannes, M., and Polson, N. (2007). MCMC maximum likelihood for latent state models. *Journal of Econometrics*, 137(2):615–640.

Jeffreys, H. (1939). *Theory of Probability.* The Clarendon Press, Oxford.

Jelinek, F. (1999). *Statistical Methods for Speech Recognition.* The MIT Press, Cambridge, MA.

Johnson, R. and Wichern, D. (1988). *Applied Multivariate Statistical Analysis*, second Edition. Prentice-Hall, Englewood Cliffs, NJ.

Jones, G., Haran, M., Caffo, B., and Neath, R. (2006). Fixed-width output analysis for Markov Chain Monte Carlo. *Journal of the American Statistical Association*, 101(476):1537–1547.

Kallenberg, O. (2002). *Foundations of Modern Probability.* Springer–Verlag, New York.

Kendall, W., Marin, J.-M., and Robert, C. (2007). Confidence bands for Brownian motion and applications to Monte Carlo simulations. *Statistics and Computing*, 17(1):1–10.

Kirkpatrick, S., Gelatt, C., and Vecchi, M. (1983). Optimization by simulated annealing. *Science*, 220:671–680.

Kong, A., McCullagh, P., Meng, X.-L., Nicolae, D., and Tan, Z. (2003). A theory of statistical models for Monte Carlo integration (with discussion). *Journal of the Royal Statistical Society Series B*, 65(3):585–618.

Lehmann, E. and Casella, G. (1998). *Theory of Point Estimation,* revised edition. Springer–Verlag, New York.

Lele, S., Dennis, B., and Lutscher, F. (2007). Data cloning: easy maximum likelihood estimation for complex ecological models using Bayesian Markov chain Monte Carlo methods. *Ecology Letters*, 10:551–563.

Liu, J. (1996). Metropolized independent sampling with comparisons to rejection sampling and importance sampling. *Statistics and Computing*, 6:113–119.

Liu, J., Wong, W., and Kong, A. (1994). Covariance structure of the Gibbs sampler with applications to the comparisons of estimators and sampling schemes. *Biometrika*, 81:27–40.

Lunn, D., Thomas, A., Best, N., and Spiegelhalter, D. (2000). WinBUGS – a Bayesian modelling framework: concepts, structure, and extensibility. *Statistics and Computing*, 10:325–337.

Marin, J.-M. and Robert, C. (2007). *Bayesian Core*. Springer–Verlag, New York.

Matthews, P. (1993). A slowly mixing Markov chain with implications for Gibbs sampling. *Statistics and Probability Letters*, 17:231–236.

McCullagh, P. and Nelder, J. (1989). *Generalized Linear Models*. Chapman and Hall, New York.

McCulloch, C. (1997). Maximum likelihood algorithms for generalized linear mixed models. *Journal of the American Statistical Association*, 92:162–170.

Metropolis, N., Rosenbluth, A., Rosenbluth, M., Teller, A., and Teller, E. (1953). Equations of state calculations by fast computing machines. *Journal of Chemical Physics*, 21(6):1087–1092.

Meyn, S. and Tweedie, R. (1993). *Markov Chains and Stochastic Stability*. Springer–Verlag, New York.

Murrell, P. (2005). *R Graphics*. Lecture Notes in Statistics. Chapman and Hall, New York.

Neal, R. (1999). *Bayesian Learning for Neural Networks*, volume 118 of *Lecture Notes in Statistics*. Springer–Verlag, New York, New York.

Neal, R. (2003). Slice sampling (with discussion). *Annals Statistics*, 31:705–767.

Newton, M. and Raftery, A. (1994). Approximate Bayesian inference by the weighted likelihood bootstrap (with discussion). *Journal of the Royal Statistical Society Series B*, 56:1–48.

Ó Ruanaidh, J. and Fitzgerald, W. (1996). *Numerical Bayesian Methods Applied to Signal Processing.* Springer–Verlag, New York.

Owen, A. and Zhou, Y. (2000). Safe and effective importance sampling. *Journal of the American Statistical Association*, 95:135–143.

Peskun, P. (1973). Optimum Monte Carlo sampling using Markov chains. *Biometrika*, 60:607–612.

Plummer, M., Best, N., Cowles, K., and Vines, K. (2006). CODA: convergence diagnosis and output analysis for MCMC. *R News*, 6(1):7–11.

Pritchard, J., Seielstad, M., Perez-Lezaun, A., and Feldman, M. (1999). Population growth of human Y chromosomes: a study of Y chromosome microsatellites. *Molecular Biology and Evolution*, 16:1791–1798.

Raftery, A. and Lewis, S. (1992). How many iterations in the Gibbs sampler? In Bernardo, J., Berger, J., Dawid, A., and Smith, A., editors, *Bayesian Statistics 4*, pages 763–773. Oxford University Press, Oxford.

Robert, C. (1993). Prior feedback: a Bayesian approach to maximum likelihood estimation. *Journal of Computational Statistics*, 8:279–294.

Robert, C. (1995a). Convergence control techniques for MCMC algorithms. *Statistical Science*, 10(3):231–253.

Robert, C. (1995b). Simulation of truncated Normal variables. *Statistics and Computing*, 5:121–125.

Robert, C. (2001). *The Bayesian Choice*, second edition. Springer–Verlag, New York.

Robert, C. and Casella, G. (2004). *Monte Carlo Statistical Methods*, second edition. Springer–Verlag, New York.

Roberts, G., Gelman, A., and Gilks, W. (1997). Weak convergence and optimal scaling of random walk Metropolis algorithms. *Annls of Applied Probability*, 7:110–120.

Roberts, G. and Rosenthal, J. (1998). Markov chain Monte Carlo: some practical implications of theoretical results (with discussion). *Canadian Journal of Statistics*, 26:5–32.

Roberts, G. and Rosenthal, J. (2009). Examples of adaptive mcmc. *Journal of Computational and Graphical Statistics*, 18(2):349–367.

Rosenthal, J. (2007). Amcm: an R interface for adaptive MCMC. *Computional Statistics and Data Analysis*, 51:5467–5470.

Rubinstein, R. (1981). *Simulation and the Monte Carlo Method.* John Wiley, New York.

Smith, A. and Gelfand, A. (1992). Bayesian statistics without tears: a sampling-resampling perspective. *The American Statistician*, 46:84–88.

Spall, J. C. (2003). *Introduction to Stochastic Search and Optimization.* John Wiley, New York.

Spector, P. (2009). *Data Manipulation with R.* Springer–Verlag, New York.

Stigler, S. (1986). *The History of Statistics.* Belknap, Cambridge, MA.

Strawderman, R. (1996). Discussion of Casella's article. *TEST*, 5:325–329.

Tanner, M. (1996). *Tools for Statistical Inference: Observed Data and Data Augmentation Methods,* third edition. Springer–Verlag, New York.

Tanner, M. and Wong, W. (1987). The calculation of posterior distributions by data augmentation. *Journal of the American Statistical Association*, 82:528–550.

Thisted, R. (1988). *Elements of Statistical Computing: Numerical Computation*. Chapman and Hall, New York.

Tufte, E. (1990). *Envisioning Information*. Graphics Press, Cheshire, CT.

Tufte, E. (2001). *The Visual Display of Quantitative Information*, second edition. Graphics Press, Cheshire, CT.

Van Laarhoven, P. and Aarts, E. (1987). *Simulated Annealing: Theory and Applications*, CWI Tract 51. Reidel, Amsterdam.

Venables, W. and Ripley, B. (1999). *Modern Applied Statistics with S-PLUS*, third edition. Springer–Verlag, New York, New York.

Wei, G. and Tanner, M. (1990). A Monte Carlo implementation of the EM algorithm and the poor man's data augmentation algorithm. *Journal of the American Statistical Association*, 85:699–704.

Wu, C. (1983). On the convergence properties of the EM algorithm. *Annals Statistics*, 11:95–103.

Zellner, A. (1986). On assessing prior distributions and Bayesian regression analysis with *g*-prior distribution regression using Bayesian variable selection. In *Bayesian Inference and Decision Techniques: Essays in Honor of Bruno de Finetti*, pages 233–243. North-Holland/Elsevier, Amsterdam.

Index of R Terms

!, 34
!=, 34
#, 5, 34
$, 13
%*%, 8
&, 34
&&, 34
|, 34
||, 34
:, 9
<-, 7
=, 7
==, 6, 34
?, 5
%%, 7, 115
%\%, 7

abline, 71
acf, 22, 43, 176
acfplot(coda), 242
all.equal, 38
amcmc, 237, 263–267
amcmc, 3
anova, 20
apply, 10, 12, 131
area(MASS), 62
argument, 12
arima, 22
arm, 20
as.matrix, 62
as.numeric, 68, 160
as.data.frame, 14
assign, 37

attach, 16, 37

backsolve, 11
barplot, 30
BATCH, 36
bayesm, 60
binomial, 21
boxplot, 30
browser, 33
BUGS, 3
bw.nrd0, 262
byrow, 9

c, 6
capabilities, 27
cars, 179
cbind, 85
cex, 28
challenger, 196
chol, 11, 73
chol2inv, 11
coda, 237, 238, 241, 248, 252, 255
coda.options(coda), 242
codamenu(coda), 242
colnames, 10
colors, 28
colours, 28
colSums, 12
contour, 31, 72, 128
cor.test, 16
crossprod, 10
cummax, 131
cumsum, 30, 66
curve, 66, 134

dbeta, 53
debug, 33
demo, 5
density, 18, 252
dev.copy, 27
dev.list, 27
dev.off, 27
dev.print, 27
dev.set, 27
diag, 10, 13
display, 20
dnbinom, 50
dnorm, 14
download.package, 5
dump, 35

each(rep), 9
effectiveSize(coda), 255
eigen, 11, 13
Energy, 204
expression, 28, 77

factor, 12
FALSE, 5
family, 21
for, 33
foreign, 35
format, 68, 145
forwardsolve, 11
function, 31

gamma, 62
gcc, 264
gelman.diag(coda), 253
gelman.plot(coda), 253
geweke.diag(coda), 248
geweke.plot(coda), 248
glm, 21, 85
    binomial, 84, 120
    quasi, 22

heidel.diag(coda), 248
heidel.diag, 247
help, 2, 9
help.search, 2
help.start, 2
hist, 27, 30

identical, 38

identify, 28
if, 33, 34
ifelse, 34, 197
image, 26, 31, 72
Inf, 62, 63
install.package, 5
integrate, 62, 65, 83, 114
is.na, 38, 261
is.vector, 8

jitter, 17
jpeg, 27

kruskal.test, 17
ks.test, 176, 245
ks.test, 17

lapply, 13
ldeaths, 22
legend, 28
length, 8
length.out(rep), 9
levelplot(coda), 242
levels, 12
lgamma, 62
library, 5
lines, 28, 30
list, 13
lm, 15, 118
locator, 28
loess(stats), 18
lty, 28
lwd, 28

mar, 27
MASS, 58, 84
matrix, 9, 10, 16
mcmc(coda), 242
mcmc.list(coda), 242, 252
mean, 31
median, 15, 31
mfrow, 27
mnormt, 48
multimenu(coda), 241

NA, 6, 98
names, 9
ncol, 9
nlm, 127

nrow, 9
ns(splines), 18
NULL, 5

objects, 36
OpenBugs, 3
optimise, 53, 127, 131
optimize, 53, 127
Orange, 38
order, 9
outer, 73, 134, 135

pairs, 31
par, 27
pbinom, 48
pch, 28
pdf, 27
permn, 8
persp, 135, 194
Pima, 21, 84, 119
    Pima.tr, 186
plot, 26, 27
plot.mcmc(coda), 252
plotmath(grDevices), 28
png, 27
pnorm, 14
points, 30, 72
    warning, 31
polygon, 30, 77
postscript, 27
ppois, 49
prod, 85

q, 36
qnorm, 14, 47
qqmath(coda), 242
qr, 11
quantile, 15, 91
quasi, 22
quit, 36

R, 2–39
    help, 2
    interface with other languages, 3,
        264
randu, 43
range, 31, 97
rank, 9
rbeta, 53

rbind, 11
rcauchy, 45
rchisq, 46, 107
.RData, 37
.RData, 5, 36
read.coda(coda), 241
read.spss(foreign), 35
read.table, 14, 35
rev, 31
rexp, 44
rgamma, 42
.Rhistory, 36
rlogis, 45
rm, 36
rmnorm(mnormt), 48, 260
rmvnorm(mvtnorm), 260
rnbinom, 50
rnegbin, 58
RNG, 42, 47
rnorm, 14
rowMeans, 12
rownames, 10
rowSums, 12
rpois, 49
rt, 50
rtnorm, 235
rtrun, 60
runif, 14, 42

sadmnv(mnormt), 48
sample, 8, 10
sapply, 13
save, 35
scan, 34, 35
sd, 15
set.seed, 43
seq, 9
shapiro.test, 17
solve, 11
sort, 9
source, 33
span(loess), 18
spline, 18
splinefun, 18
splines, 18
str, 5
sum, 8
summary, 18, 118
summary.mcmc(coda), 252

svd, 11
Sweave, 2
swiss, 190
switch, 34, 231
system, 33
system.time, 7, 34, 38, 46

t, 8
table, 12
tapply, 12
tcrossprod, 11
times(rep), 9
title, 28, 145
trace, 33
TRUE, 5
t.test, 16

unique, 143
uniroot, 160, 161

var, 15, 95
vi, 33

warnings, 10
while, 33, 54
wilcox.test, 17
WinBugs, 3

X11, 27
xlim, 31
xor, 34

ylim, 31

Index of Subjects

ABC, 56
Accept–Reject, 113, 115, 176
    and recycling, 124
    criticism, 57
    tight bound, 55
acceptance probability, 53
acceptance rate, 167, 171, 183
    optimal, 195
adaptive algorithms, 258
    diminishing adaptation, 263
algorithm
    acceleration of, 111
    Box–Muller, 47
    comparison, 111
    EM, 152, 215
    greedy, 91
    independent Metropolis–Hastings,
        176
    Langevin , 186
    MCEM, 157
    MCMC, 169
    Metropolis–Hastings, see
        Metropolis–Hastings algo-
        rithm
    optimization, 55, 111
    random walk Metropolis–Hastings,
        182
    simulated annealing, 140
allocation indicator, 164
analysis of variance, 20, 253
analysis, spectral, 248
approximate Bayesian computation
        (ABC), 56

approximation, normal, 255
autocorrelation, 22

batch sampling, 240
    batch size, 245
Bayes factor computation, 87, 93, 229
Bayesian inference and decision theory,
        62
Boltzman–Gibbs transform, 140
bootstrap, 23, 77, 90, 91
Box-Muller algorithm, 47
Brownian motion, 103

calibration, 267
Challenger data, 196
Chib's approximation, 230
coda convergence assessment, 238
Comprehensive R Archive Network
        (CRAN), 3, 264
computing time, 34, 56, 91, 176, 195,
        232, 266
conditional distributions, 204
conditioning, 107, 108, 227
confidence band, 89, 90
continuity, 94
control variate, 116, 119, 120
convergence
    acceleration, 133
    assessment, 111
        multivariate, 96
        of variability, 91
    graphical evaluation of, 242
    monitoring of, 67, 90

of Markov chain Monte Carlo
    algorithms, *see* MCMC
    algorithm
    slow, 81
    test, 65
    to the stationary distribution, 239
convergence
    assessment, 281
correlation, 111, 113
covariance matrix, 73
Cramer–von Mises test, *see* test

data
    censored, 153
    complete, 151
    missing, 151
data augmentation, 158
decision theory, 62
defensive sampling, 81
demarginalization, 146, 151
density
    instrumental, 51
    spectral, 247
    target, 51
detailed balance condition, 172
Dickey's decomposition , 107
distribution
    beta-binomial, 122, 202
    conditional, 232
    conjugate, 119
    discrete, 48
    empirical, 23
    gamma, 58
    Gumbel, 88
    instrumental, 51, 56, 71, 171, 176
    invariant, 169
    inverse Gaussian, 193
    inverted gamma, 202
    normal, 47, 114, 183
    Pareto, 81
    t, 80
    target, 170
    truncated normal, 59
    uniform, 42, 115
    witch's hat, 267
division by zero, 98
DNA sequence, 143
dyadic
    sequence, 114

symmetry, 114

effective sample size, 77, 85, 99, 195,
    256
eigenanalysis, 11
EM algorithm, 152
    monotonicity, 153
    Monte Carlo, 157
    steps, 153
empirical cdf, 243
empirical mean, 65, 108, 115, 122, 192,
    204, 227, 228, 255
    and Rao–Blackwellization, 235
    convergence, 240
Ergodic Theorem, 169
ergodicity, 169
estimator
    James–Stein
        truncated, 112
    path sampling, 94
    Rao–Blackwellized, 123
exponential family, 120
exponential variate generation, 44, 46

fixed point, 153
Fourier analysis, *see* spectral
full conditionals, 206, 235
function
    log-cumulant, 120
    loss, 111

generalized linear models, 21
generator, pseudo-random, 42
genetic linkage, 158
Gibbs random field, 200
Gibbs sampling, 200–234
    completion, 209
    definition, 206
    two-stage, 200–204
gradient, 127, 137
    stochastic, 137

Hessian, 127
hybrid strategies, 231

importance sampling, 69, 149
    accuracy, 71
    by defensive mixtures, 82
    efficiency, 81

identity, 70
instrumental distributions, 69
multiple, 82
principle, 69
self-normalized, 76, 93, 98
variance, 95
weight, 75
weight, normalized, 99
independence, 241
inference
asymptotic, 62
statistical, 62
initial distribution
parallel chains, 239
integer part, 86
integration, 62
approximative, 65
Monte Carlo, 65, 67
numerical bounds, 62
irreducibility, 169

kernel, 168
Kolmogorov–Smirnov statistic, 245
Kullback–Leibler divergence, 99

Langevin algorithm, 186
Law of Large Numbers
extension, 169
strong, 65
linear models, 14, 18
generalized, 21
Linux, 3, 241
local maxima, 133
log-cumulant function, 120
logistic regression, 219

marginalization, 82
Monte Carlo, 88
Markov chain, 137, 167
adaptive algorithms, 258
empirical cdf, 243
Ergodic Theorem, 169
limiting distribution, 169
local exploration, 182
parallel chains, 239
stationary distribution, 169
transition kernel, 168
maxima, local, 137
MCEM algorithm, 157, 161

MCMC algorithm
convergence of, 238
monitoring of, 238
measurability, 94
method
Accept–Reject, 51, 56
gradient, 137
kernel, 247
Monte Carlo, see Monte Carlo
Newton–Raphson, 127, 137
numerical, 126
Metropolis within Gibbs, 230
Metropolis–Hastings algorithm, 141, 232
independent, 176
probability of acceptance, 171
random walk, 193
missing data, 146, 209
missing mass, 240
mixture
defensive, 82
exponential, 163
indicator, 214
normal, 154
for simulation, 50, 81, 107
stabilization by, 81
MLE (maximum likelihood estimator), 85
mode, 132
model
ARMA, 247
augmented, 158
censored-data, 210
choice, 93
completed, 215
generalized linear, 15, 21
hierarchical, 221
logistic, 119, 219
missing-data, 210
mixture, 214
multinomial, 211
probit, 83, 147, 186
random effects, 233
model choice, 188
monitoring of Markov chain Monte
Carlo algorithms, see MCMC
algorithm
monotonicity of covariance, 241
Monte Carlo, 42, 69, 70, 81, 112

approximation, 147
EM, 157, 159
hybrid, 231
marginalization, 82, 88
validation, 112

∇ (gradient), 127
Newton–Raphson, 127
non-parametric statistics, 14
nonstationarity, 242
normal variate generation, 47
normalizing constant, 93, 175
nuclear pump failures, 222, 245
numerical integration, 62

O-ring, 196
optimization, 56, 62

parallel chains, 239
path, 239
perplexity, 99
Poisson process, 222
prior
    feedback, 141
    improper, 232
probability
    distributions in R, 14
    integral transform, 44
probability of acceptance, *see* acceptance probability

R, 2–39
    Box–Muller random generator, 47
    CRAN, 3
    data, 36
    data frame, 14
    debugging, 33
    depository, 31
    factor, 12
    functions, 31
    history, 36
    list, 13
    matrix, 10, 13
    probability distributions, 14
    programming, 31
    vector, 7
    versus BUGS, 3
random effects, 233
random variables

antithetic, 113
beta, 58
Cauchy, 45
discrete, 48
exponential, 44, 46
gamma, 58
logistic, 45
mixture, 50
multivariate normal, 47
negative binomial, 50
noncentral chi-squared, 59
normal, 47
Pareto, 57
truncated normal, 59, 235
uniform, 42, 45, 57
random walk, 169, 193, 194
randu, 57
Rao–Blackwellization, 107–111, 227
    implementation, 108
    termwise, 123
rate of acceptance, 194
raw sequence plot, 252
recurrence, 169
recycling, 227
regeneration, 91
regression
    linear, 254
    logistic, 119
resampling
    and bootstrap, 92
    degeneracy, 77
    multinomial, 75
    unbiased, 87
running mean plot, 96, 100

saddlepoint, 137
SAME algorithm, 141
sample
    independent, 111
    uniform, 112
sample size, effective, *see* effective sample size
sampling
    defensive, 81, 110
    importance, 253
    stratified, 110
shrink factor, 253
shuttle Challenger, 196
simulated annealing, 182

temperature schedule, 133, 140,
    141
simulation, 62, 232
    in parallel, 91
    univariate, 206
    versus numerical methods, 126
slice sampler, 216
spectral analysis, 247
speed of convergence, 194
stability of a path, 81
stationarity, 239, 240, 242–250, 252
stationary distribution, 169
    as limiting distribution, 169
stochastic
    approximation, 130
    optimization, 174
    search, 130
stochastic gradient, 182
stopping rule, 253
stopping time, 238
Student's t variate generation, 50
subsampling, 240, 245
    and convergence assessment, 241
    and independence, 245
sudoku, 39
sufficient statistic, 107
support, 51, 65, 175

tail probability estimation, 70
temperature schedule, 133, 140, 141
termwise Rao–Blackwellized estimator,
    123
test
    Cramer–von Mises, 245
    halfwidth, 248
    Kolmogorov–Smirnov, 173, 245
    likelihood ratio, 69

non-parametric, 245
power of, 69
stationarity, 239
testing, 69
theorem
    Central Limit, 62, 65, 90
    Donsker, 103
    Ergodic, 169, 240
    fundamental, simulation, 52
    Hammersley–Clifford, 233
    Rao–Blackwell, 107
transition kernel, 168
traveling salesman problem, 126, 143

uniform random variable
    generation, 45

variable
    antithetic, 111, 113
    auxiliary, 209
    control, 120
    latent, 215
variance
    between- and within-chains, 253
    finite, 79
    of a ratio, 93
    reduction, 107, 108, 112, 124
        and Accept–Reject, 115
        and antithetic variables, 123
        and control variates, 116, 119
        optimal, 111
variate (control), 116

You've only seen where you've been,
    239, 247, 249

Z-score, 248

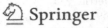 **Springer**   **springer.com**

the language of science

Monte Carlo Statistical Methods

Christian P. Robert
George Casella

This new edition has been revised towards a coherent and flowing coverage of these simulation techniques, with incorporation of the most recent developments in the field. In particular, the introductory coverage of random variable generation has been totally revised, with many concepts being unified through a fundamental theorem of simulation.

Content: Introduction.- Random Variable Generation.- Monte Carlo Integration.- Markov Chains.- Monte Carlo Optimization.- The Metropolis-Hastings Algorithm.- The Gibbs Sampler.- Diagnosing Convergence.- Implementation in Missing Data Models.

2005. 2nd ed. 2004. Corr. 2nd printing XXX, 645 p. 132 illus. Hardcover
Springer Texts in Statistics
ISBN: 978-0-387-21239-5

Bayesian Core: A Practical Approach to Computational Bayesian Statistics

Jean-Michel Marin
Christian P. Robert

Content: User's manual.- Normal models.- Regression and variable selection. - Generalised linear models.- Capture-recapture experiments.- Mixture models.- Dynamic models.- Image analysis.

2008. 1st ed. 2007. Corr. 2nd printing XIV, 258 p. 80 illus. Hardcover
Springer Texts in Statistics
ISBN: 978-0-387-38979-0

Bayesian Computation with R

Jim Albert

Content: An introduction to R.- Introduction to Bayesian thinking.- Single-parameter models.- Multiparameter models.- Introduction to Bayesian computation.- Markov chain Monte Carlo methods.- Hierarchical modeling.- Model comparison.- Regression models.- Gibbs sampling.- Using R to interface with WinBUGS.

2009. 2nd ed. XII, 300 p. Softcover
Use R
ISBN: 978-0-387-92297-3

Easy Ways to Order ▶   Call: Toll-Free 1-800-SPRINGER • E-mail: orders-ny@springer.com • Write: Springer, Dept. S8113, PO Box 2485, Secaucus, NJ 07096-2485 • Visit: Your local scientific bookstore or urge your librarian to order.